普通高等教育能源动力类系列教材

制冷空调专业英语

孙　晗　编著

机　械　工　业　出　版　社

本书内容包括制冷空调科技英语阅读和写作两部分。阅读部分有课文16篇，涉及传热学、热力学、流体力学、空调制冷等方面的内容。每篇课文后有词汇表、学习要点和练习。学习要点主要介绍与课文相关的重要语法知识。写作部分着重介绍学术英语的写作要点，并配有相关练习。全书取材广泛，除扩大学生专业词汇量外，重在培养学生制冷空调科技英语的阅读和写作能力。

本书为能源与动力工程专业制冷空调方向本科生的专业英语教材，也可供建筑环境与能源应用工程专业学生选用，可作为研究生的科技英语阅读和写作教材，还可供制冷空调领域有科技英语阅读和写作需求的工程技术人员自学参考。

图书在版编目（CIP）数据

制冷空调专业英语/孙晗编著. —北京：机械工业出版社，2021.1（2025.1重印）

普通高等教育能源动力类系列教材

ISBN 978-7-111-67221-0

Ⅰ.①制… Ⅱ.①孙… Ⅲ.①制冷-空气调节设备-英语-高等学校-教材
Ⅳ.①TB657.2

中国版本图书馆CIP数据核字（2020）第272471号

机械工业出版社（北京市百万庄大街22号 邮政编码100037）
策划编辑：蔡开颖 责任编辑：蔡开颖 尹法欣
责任校对：段晓雅 封面设计：张 静
责任印制：张 博
北京雁林吉兆印刷有限公司印刷
2025年1月第1版第2次印刷
184mm×260mm · 7.5印张 · 166千字
标准书号：ISBN 978-7-111-67221-0
定价：24.80元

电话服务
客服电话：010-88361066
010-88379833
010-68326294

网络服务
机 工 官 网：www.cmpbook.com
机 工 官 博：weibo.com/cmp1952
金 书 网：www.golden-book.com
机工教育服务网：www.cmpedu.com

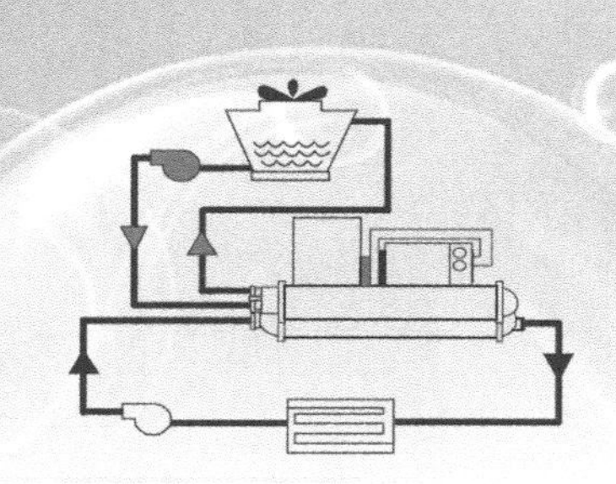

前言

现代社会已离不开制冷空调。制冷空调技术目前已广泛应用在日常生产、生活和科学研究的方方面面。英文是世界各国的暖通、空调、制冷工程师进行技术交流的主要语言。英文论文、专利、标准、教材等是制冷空调领域的研究人员、工程师、教师及学生了解行业发展动态、跟踪学科前沿、学习制冷新技术新理论的重要途径。国际英文期刊刊登世界各国各领域专家学者的原创研究论文和综述论文，经常阅读英文期刊可提高专业技术水平，启发创新思维。专业英文期刊的读者往往也是英文期刊论文的撰写者，因此在阅读期刊论文获取知识的同时，也需要学习撰写期刊论文的方法。

大量阅读英文期刊论文是学习撰写学术论文的主要途径。撰写英文期刊论文除应了解科技论文写作的特点外，还需要掌握大量专业领域的专业词汇。具有大学英语四、六级水平但从未接触过专业英语的学生直接阅读英文文献，初期也会感觉吃力，有些专业词汇可能只知道它的一个常用意思，但不知道在专业领域该如何正确翻译。例如，知道coil是线圈的意思，但不知道fan coil为风机盘管。直接阅读英文文献可能会使部分学生产生较强挫败感，从而失去阅读的兴趣。为了使学生从开始阅读吃力过渡到无障碍阅读英文文献，有必要先让学生读一些内容较简单的专业短文，使学生逐步扩大专业词汇量，然后再阅读英文原版教材和期刊论文，这样使学生逐渐适应英文期刊论文的难度，较顺利地阅读。阅读达到既快速又准确后，撰写论文就变得容易了。本书正是基于这一理念编写的。

本书共两部分。第一部分为阅读部分，以训练学生阅读能力为主，包括课文16篇，通过学习课文掌握传热学、热力学、流体力学、空调制冷等方面的基本专业词汇。每篇课文后有词汇表、学习要点和练习等相关内容。其中，学习要点主要是介绍与课文内容相关的语法要点和写作技巧提示；练习部分主要有翻译、语法要点练习等，并留有空行，方便学生在书上做练习或记录练习答案。本书配套的课件中提供课文翻译及习题答案。第二部分为写作部分，以介绍学术论文的写作特点为主，介绍期刊原创研究论文的一般结构，说明各部分的写作要点，并给出训练论文写作的一些练习。

由于作者水平有限，书中难免存在不足之处，恳请读者批评指正。

感谢北京市委组织部优秀人才项目提供的出版资助。感谢清华大学石文星教授、北京工业大学刘忠宝教授在教材撰写初期对教材内容提出的宝贵建议。感谢家人给予我的全力支持和鼓励。

孙　晗

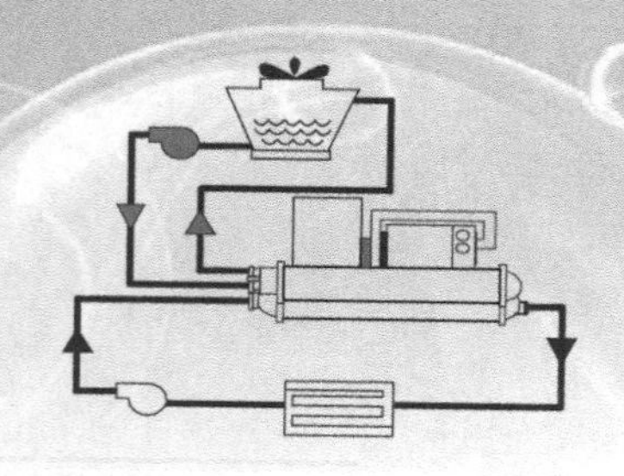

目录

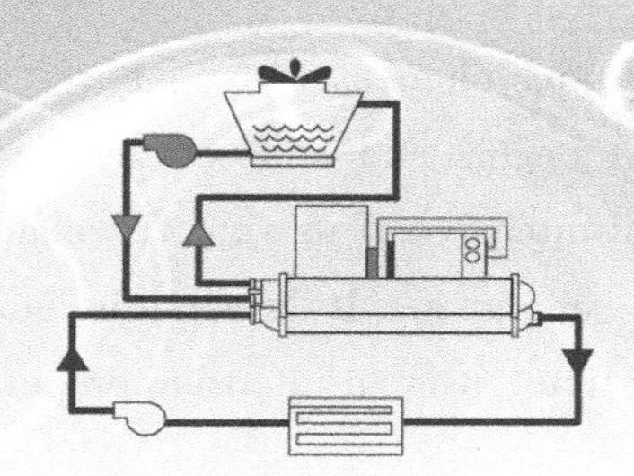

第1部分

制冷空调科技英语阅读

1.1 Basic Concepts of Thermodynamics

The word thermodynamics stems from the Greek words therme (heat) and dynamis (force). Thermodynamics studies basic laws of thermal processes, thermophysical properties of matter and working processes of various thermal equipment.

The term system is used to identify the subject of the analysis. Everything external to the system is considered to be the system's surroundings. The system is distinguished from its surroundings by a specified boundary.

The closed and open systems are two basic kinds of systems. A closed system is a system in which a certain quantity of matter is under study. There is no transfer of mass across its boundary. A special type of closed system that does not interact in any way with its surroundings is called an isolated system. An open system is a system where mass may cross the boundary of a control volume.

The closed and open system can also be called control mass and control volume respectively. When the terms control mass and control volume are used, the system boundary is often called a control surface.

Thermodynamics can be classified into classical and statistical thermodynamics. Classical thermodynamics studies systems with the macroscopic approach, while statistical thermodynamics studies systems from a microscopic point of view. For most engineering applications, classical thermodynamics is adopted to analyze the real problems. To promote understanding, the behavior of the system can be explained with statistical thermodynamics.

Properties are macroscopic characteristics of a system such as temperature, pressure, density, specific capacity, specific enthalpy, viscosity and surface tension. The term state means the condition of a system as described by its properties. A process is a transformation from one state to another. A system is said to be at steady state if none of its properties changes with time.

A thermodynamic cycle includes a sequence of processes that begins and ends at the same state. The properties can be divided into two classes: extensive and intensive. If its value is related to mass or size, it is an extensive property such as mass, volume, energy etc. If its value is independent of mass or size and may be functions of both position and time, it is an intensive property such as temperature and pressure.

The word phase refers to a state of matter that is homogeneous in both chemical composition and physical structure under the specific pressure and temperature. The phase of matter can be solid, liquid and vapor. A system can contain one or more phases.

A pure substance is one that is uniform and invariable in chemical composition. For example, liquid water and water vapor from a system are two phases but can be regarded as a pure substance.

Equilibrium is a fundamental concept in thermodynamics. Equilibrium is a condition of balance maintained by the equality of opposing influences. A system in equilibrium means that there is no tendency to change its condition when the overall system is isolated from its surroundings. The complete equilibrium includes mechanical, thermal, phase and chemical equilibrium.

A quasi-equilibrium process is an idealized process in which the departure from thermodynamic equilibrium is at most infinitesimal. Considering that nonequilibrium is inevitable during actual processes, a real process can be regarded as compositions of many quasi-equilibrium processes. All states through which the system passes in a quasi-equilibrium process may be considered as equilibrium states.

词汇表

thermodynamics 热力学

stem from 起源于

thermophysical property 热物理性质（热物性）

system 系统

surroundings 环境，外界

term 术语

identify 确定，识别

subject 主体

external 外部的

be distinguished from 区别于

specified 指定的，特定的

boundary 边界

a closed system 闭口系统

an open system 开口系统
matter 物质
an isolated system 孤立系统
mass 质量，物质
be classified into 被划分为
classical thermodynamics 经典热力学
statistical thermodynamics 统计热力学
macroscopic 宏观的
microscopic 微观的
approach 方法
property 性质
temperature 温度
pressure 压力
density 密度
enthalpy 焓
specific capacity 比热容
specific enthalpy 比焓
viscosity 黏度
surface tension 表面张力
state 状态
process 过程
steady state 稳态
cycle 循环
extensive property 广度量
intensive property 强度量
volume 体积，容积
energy 能量，能源
function 函数
phase 相
homogeneous 均匀一致的
physical structure 物理结构
equilibrium 平衡
fundamental 基本的
concept 概念
balance 平衡
equality 相等

mechanical equilibrium　力平衡
thermal equilibrium　热平衡
phase equilibrium　相平衡
chemical equilibrium　化学平衡
idealized　理想化的
departure　离开
infinitesimal　无限小的，无穷小的
inevitable　不可避免的

学习要点

科技论文经常使用结构复杂的长句。有时一个长句中每个单词都认识，但就是读不懂整个句子的意思。为了克服此类阅读障碍，就必须主动学习英语语法知识。阅读时，掌握英语语法知识有助于理解复杂的长句。写作时，掌握语法知识，可使写出的句子正确、流畅。

1. 句法和词法

语法（grammar）包括句法（syntax）和词法（morphology）。

例 1.1-1　Experimental results show that adding a small amount surfactant into lithium bromide aqueous solution can increase absorption rate obviously.

实验结果表明向溴化锂水溶液中添加微量表面活性剂可显著提高吸收率。

该例句可分为三个主要部分（items）。第一部分 experimental results，第二部分 show，第三部分 that 从句。其中，experimental results 是主语（subject），动词 show 是谓语（predicate），谓语后面是 that 引出的宾语从句（object clause）。这个句子是陈述句。主语在谓语前，宾语从句在谓语后。句中这三个主要部分的位置按照主谓宾这样的顺序就是正确的。如果句子仍为陈述句，这三者位置发生改变，语法上就是错的。句中的词、词组及从句的位置关系属于语法中句法的范畴。

Results 是 result 的复数形式。该例句时态是一般现在时，动词为 show，show 不加 s 就是对的，因为前面的 results 是复数，这些属于语法中的词法。词法是关于词型变化的，如名词的数、格，动词的时态、语态。

2. 句子和从句

句子（sentence）是由至少一个主句（main clause）及从句（subordinate clause）组成的。一个句子至少包含一个主句。无论主句还是从句都包含两个最重要的句子成分——主语和谓语。名词词组或相当于名词词组的短语可以作主语，一个从句也可以作主语，表示动作的执行者、状态或事件。谓语是动词词组和其他伴随的内容。除了主语和谓语，句子中有时还有表语、宾语、定语、状语等成分。主句可独立存在，从句依赖主句而存在。

例 1.1-2　When water temperature reaches the set point, the electrical heater switches off automatically.

该例句包含两个子句（clause）。前面的子句是从句，后面的子句是主句。从句中water temperature是主语，reach是谓语。主句中the electrical heater是主语，switches off是谓语。注意主句不依赖其他从句，本身就是一个完整的句子。而when water temperature reaches the set point是从句，是时间状语从句，它依赖主句的存在而存在，它不能单独存在。

3. 简单句、并列句和复合句

一个句子只有一个主句就是简单句。

例1.1-3 Temperature difference is the driving force of heat transfer.

一个句子如果有一个主语，两个或多个谓语，也是简单句。

例1.1-4 A new experimental setup was designed and built.

该例句有一个主语a new experimental setup，两个谓语分别是was designed和was built，这样的句子因为只有一个主语也是简单句。

两个或更多的主句构成的句子就是并列句。

例1.1-5 Vapor compression water chiller is powered by electricity and absorption water chiller is driven by heat.

该例句中两个主句通过连词and构成一个并列句。两个主句都有主语和谓语。第一个主句主语是vapor compression water chiller，谓语是is powered；另一个主句主语是absorption water chiller，谓语是is driven。

一个句子由一个主句和一个或多个从句组成，就是复合句。

例1.1-6 If a small amount of nanoparticles are added into solution, the absorption rate increases.

该例句中有一个主句the absorption rate increases和一个if引导的条件状语从句if a small amount of nanoparticles are added into solution，因此整个句子是一个复合句。

4. 定语从句或关系从句（relative clause）

定语从句告诉我们哪个人、哪件事或哪个地方是什么样的人、事、地方。关系从句分为限定性从句和非限定性从句。

（1）限定性从句

例1.1-7 A doctor is a person who treats patients in hospital.

该例句是用关系代词who引导的从句，用来定义doctor是什么样的人，是在医院治疗病人的人。因为从句前的名词是person，所以从句使用关系代词who来引出从句。Who在从句中为主语，不可省略。该从句为限定性从句，从句who treats patients in hospital如果省略，句子的意思就不完整。

例1.1-8 A special type of closed system that (which) does not interact in any way with its surroundings is called an isolated system.

该例句中定语从句要说明的是一种closed system。system是一个事物，可用that或which引出的从句来说明一种特殊形式的闭口系统，是这样一种系统，是一种不以任何方式和周围环境相互作用的系统，这样的系统是孤立系统。在该从句中that或which作主语，不可省略。

例 1.1-9 Have you found the bicycle that you lost?

该例句中 that 引出的从句修饰 bicycle。that 在从句中是宾语，这种情况 that 可省略。

例 1.1-10 An open system is a system where mass may cross the boundary of a control volume.

该例句中用关系副词 where 引出的从句来说明在 system 处，物质穿过控制体的边界。

以上例句均为限定性定语从句，用来定义或描述从句前面的名词，如果省略从句，主句的意思就不完整。

（2）非限定性从句

例 1.1-11 A familiar device for temperature measurement is the liquid-in-glass thermometer, which consists of a bulb filled with a liquid such as alcohol and a glass capillary tube.

该例句中 which 引导的从句是对 thermometer 的补充说明，而不是定义。去掉从句也不会对主句产生太大影响。

科技英语中下定义，经常使用定语从句。课文中有一个典型的使用定语从句为一个专业词汇下定义的句子。

例 1.1-12 A quasi-equilibrium process is an idealized process in which the departure from thermodynamic equilibrium is at most infinitesimal.

A quasi-equilibrium process 是要定义的概念，is 是下定义时常用的动词形式。其他下定义时常用的动词形式还可以使用 means/describes/may be defined as/can be defined as/is defined as。an idealized process 是定义了 A quasi-equilibrium process 是什么或属于什么范畴，in which the departure from thermodynamic equilibrium is at most infinitesimal 是 in which 引出的定语从句，具体说明是什么样的理想过程，是离开热平衡无限小的过程。这里 which 前有介词 in，这种情况介词后只能用 which，不能用 that。

5. 科技英语中分类的用法

常用的句型是：

... be divided (classified) into ...

例如课文中的句子：

Thermodynamics can be classified into classical and statistical thermodynamics.

练习

1. 将下列句子翻译成英语。

1）物性参数可分为广度量和强度量。

2）空气调节系统可分为全空气系统、空气-水系统和冷剂系统。

3）热电偶是一种由两种不同金属丝焊接而成的温度传感器。

2. 请将下列句子译成汉语。

1）The specific volume is defined as the reciprocal of density.

2）A familiar device for temperature measurement is the liquid-in-glass thermometer, which consists of a bulb filled with a liquid such as alcohol and a glass capillary tube.

3）Refrigeration is a technology which is concerned with the cooling of bodies or fluid to temperatures lower than those available in the surroundings at a particular time and place.

3. 请说明第 2 题的几个句子是简单句、并列句还是复合句。每个句子的主句在哪里？主语、谓语各是什么？有从句吗？从句中有主语和谓语吗？主语、谓语各是什么？

4. 请各写出 1~2 个简单句、并列句和复合句。

5. 请写几个复合句，并带有限定性定语从句。

6. 请写几个复合句，并带有非限定性定语从句。

7. 对话练习。

可以两人一组，一人角色为老师，一人角色为学生。老师对学生的情况并不熟悉，想了解学生的以下情况：在哪个学校上大学？是什么专业？学习了哪些主要课程？是否学过工程热力学？学生回答老师的提问，并向老师介绍工程热力学主要学习了哪些内容。

1.2 First Law of Thermodynamics

The law of energy conservation and conversion is a basic law in nature. It states that all matter have energy. Energy can neither be created nor be eliminated. It can only converts from one form to another. The total energy keeps constant in conversion.

The first law of thermodynamics is the application of law of energy conservation and conversion in thermodynamics.

For a closed system, the change of energy contained within the system during some time interval (E_2-E_1) is equal to the difference between net amount of energy transferred in across the system boundary by heat transfer Q and that out across the system boundary by work W.

The energy balance can be expressed by the following equation

$$E_2-E_1=Q-W \tag{1.2-1}$$

The plus sign before Q means that the heat transfer of energy is from the surroundings into the system and the minus sign before W means that the energy transfer by work is from the system to the surroundings.

Eq. (1.2-1) can also be written in an alternative form as

$$\Delta E_k+\Delta E_p+\Delta U=Q-W \tag{1.2-2}$$

where ΔE_k is the change in kinetic energy, ΔE_p is the change in gravitational potential energy and ΔU is the change in internal energy.

The energy balance can also be written in differential form as

$$\mathrm{d}E=\delta Q-\delta W \tag{1.2-3}$$

The time rate form of the energy balance is

$$\frac{\mathrm{d}E}{\mathrm{d}t}=\dot{Q}-\dot{W} \tag{1.2-4}$$

The energy balance takes the following form for a thermodynamic cycle

$$\Delta E_{cycle}=Q_{cycle}-W_{cycle} \tag{1.2-5}$$

where Q_{cycle} represents net amount of energy transfer by heat and W_{cycle} represents net amount of energy transfer by work for a cycle. ΔE_{cycle} is equal to zero because the system is returned to its initial state after the cycle.

Therefore Eq. (1.2-5) reduces to

$$W_{cycle}=Q_{cycle} \tag{1.2-6}$$

A pressure-enthalpy diagram of a theoretical refrigeration cycle is shown in Fig. 1.2-1.

$$W_{cycle}=h_2-h_1 \tag{1.2-7}$$

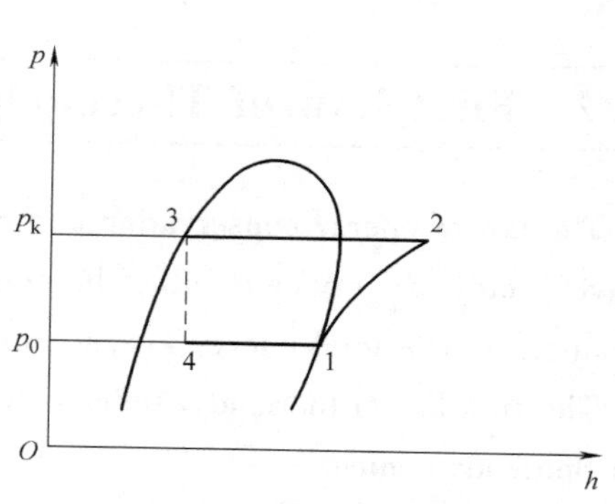

Fig. 1. 2-1 *p-h* diagram for a theoretical refrigeration cycle

where W_{cycle} is specific work of compressor, h_2 is specific enthalpy of point 2 and h_1 is specific enthalpy of point 1.

$$Q_1 = h_1 - h_4 \qquad (1.2\text{-}8)$$

where Q_1 is the heat absorbed from refrigerated space.

$$Q_2 = h_2 - h_3 \qquad (1.2\text{-}9)$$

where Q_2 is the heat released from system to surroundings.

Based on Eq. (1. 2-6),

$$Q_{cycle} = Q_2 - Q_1 = W_{cycle} \qquad (1.2\text{-}10)$$

The coefficient of performance is obtained as

$$COP = \frac{Q_1}{W} (\text{refrigeration cycle}) \qquad (1.2\text{-}11)$$

In respect that gas or liquid passes through compressors, fans or pumps, it is more convenient to use an open system to analyze them. An example of an open system is shown in Fig. 1. 2-2.

For the one-dimensional flow as shown in Fig. 1. 2-2, the energy rate balance can be written in the following form.

$$\frac{dE_{cv}}{dt} = \dot{Q} - \dot{W} + \dot{m}_i\left(u_i + \frac{v_i^2}{2} + gz_i\right) - \dot{m}_o\left(u_o + \frac{v_o^2}{2} + gz_o\right) \qquad (1.2\text{-}12)$$

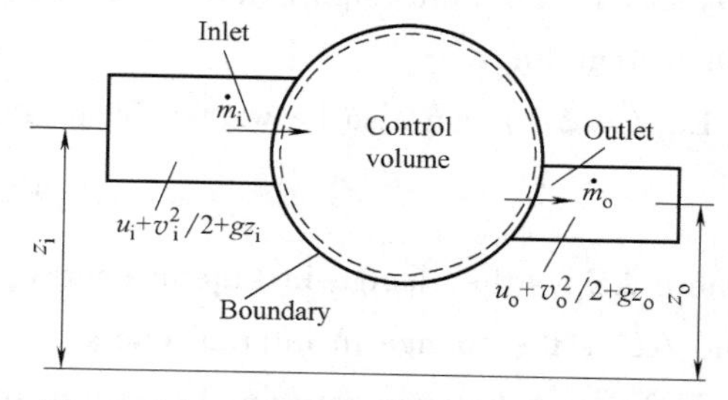

Fig. 1. 2-2 Schematic of an open system

where E_{cv} denotes the energy of the control volume at time t. The terms $\dot{Q}$ and $\dot{W}$ represent the net rate of energy transfer by heat and work across the boundary of the control volume at t respectively. $\dot{m}_i u_i$ is the internal energy of the entering stream and $\dot{m}_o u_o$ is the internal energy of the exiting stream. $\dot{m}_i \frac{v_i^2}{2}$ is the kinetic energy of the entering stream and $\dot{m}_o \frac{v_o^2}{2}$ is the kinetic energy of the exiting stream. $\dot{m}_i g z_i$ and $\dot{m}_o g z_o$ account for the potential energy of the entering stream and the exiting stream respectively. If there is no mass flow in or out, the last 2 terms on right side vanish and Eq. (1. 2-12) reduces to Eq. (1. 2-4).

词汇表

first law of thermodynamics 热力学第一定律
eliminate 消除
convert, conversion 转换

conservation　守恒
interval　间隔
work　功
net　净余的
alternative　可替代的
differential　微分的
kinetic energy　动能
gravitational potential energy　重力势能
internal energy　内能
rate　比率，速率
initial state　初始状态
coefficient of performance（COP）　性能系数
control volume　控制体
define　定义
one-dimensional flow　一维流动
account for　表示
vanish　消失

学习要点

1. 动词的限定形式

动词作谓语时，动词的形式要受主语的限制，要和主语在人称和数上保持一致。动词的形式反映出时态、语态和语气。这时的动词形式称为动词的限定形式。

1）与主语在人称和数上一致。

例 1.2-1　The law of energy conservation and conversion is a basic law in nature.

动词 is 和主语 The law 在人称和数方面是一致的。The law 是第三人称单数，系动词用 is。这里 is 就是动词 be 的限定形式。

2）动词反映出时态、语态及语气。

例 1.2-2　The energy balance can also be written in differential form as

$$dE = \delta Q - \delta W$$

该例句中 can be written 就是动词 write 的限定形式，是带有情态动词 can 的被动语态。

2. 动词的非限定形式

动词的非限定形式，不能独立作谓语。但可以和助动词或情态动词等一起构成谓语，还可在句中充当主语、宾语、定语、状语和表语等。动词的非限定形式包括动名词、分词（现在分词、过去分词）、不定式。

1）动名词：动词+ing。主要起名词的作用，可以作主语、宾语、表语等，也可用在介词后。

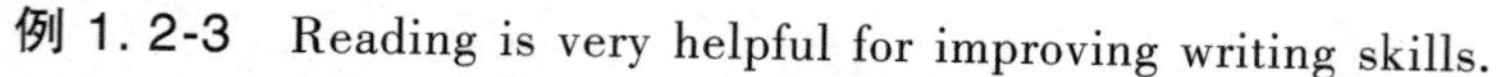

例 1.2-3 Reading is very helpful for improving writing skills.

这里 reading 就是动名词作主语。improving 用在介词 for 后面也是动名词，这里 writing 是 write 的现在分词形式，用在名词 skill 前面相当于形容词，因此这里的 writing 不是动名词。

2）现在分词：现在分词和动名词的形式相同，但不起名词的作用。可和助动词 be 一起构成现在进行时。现在分词有时也可作状语使用，现在分词作状语时，通常表示主语正在进行另一个动作，对谓语表示的动作进行修饰或陪衬。

例 1.2-4 和前面的助动词 be 构成现在进行时。

He is reading a book.

这里 reading 是现在分词，和前面的助动词 be 构成现在进行时，reading 不能单独作谓语，但 reading 和 is 在一起 is reading 是谓语。

例 1.2-5 现在分词作状语。

The cooling water flows through the channel carrying condensation heat.

这里 carrying 是现在分词作状语。carry 和 flow 动作几乎同时发生，flow 是主句的谓语，而另一个动作 carry 就可以用现在分词形式。注意主句 flow 对应的主语是 the cooling water，carry 动作对应的主语也是 the cooling water，两者的主语应保持一致。

例 1.2-6 现在分词作定语。

One of the most important factors influencing the capacity and efficiency of vapor compression refrigerating systems is the specific volume of the refrigerant vapor leaving the evaporator.

这里 influencing the capacity and efficiency 是现在分词短语作定语修饰前面的名词短语 important factors，leaving the evaporator 是现在分词短语作定语修饰前面的名词短语 the refrigerant vapor。

3）过去分词：被动语态由助动词 be 加过去分词构成。

例 1.2-7 The data is collected by a data acquisition instrument.

这里动词 collect 的过去分词是 collected，is 和 collected 一起构成被动语态，is collected 作谓语。

例 1.2-8 This is an unexpected result.

这是一个想不到的结果。

这里 unexpected 是过去分词作定语，修饰 result。

例 1.2-9 This is the method recommended by Prof. A.

该例句中过去分词短语 recommended by Prof. A 作定语，修饰前面的名词 method。此处也可以不用过去分词，而用定语从句表达。

例 1.2-10 Driven by waste heat，absorption chiller is eco-friendly.

这里的 driven by waste heat 是过去分词作状语。

4）不定式：不定式是动词的非限定形式。一般在动词原形前加 to。可以作主语、宾语、表语、定语或状语。

例 1.2-11 To promote understanding，the behavior of the system can be explained with statistical thermodynamics.

这里 to promote understanding 这个不定式短语作句子的目的状语。

3. 常用表达形式

1）……表示(代表)……

常用动词为 denote/represent/account for/mean/be 等。

例 1.2-12

$$\frac{\mathrm{d}E_{\mathrm{cv}}}{\mathrm{d}t}=\dot{Q}-\dot{W}+\dot{m}_{\mathrm{i}}\left(u_{\mathrm{i}}+\frac{v_{\mathrm{i}}^{2}}{2}+gz_{\mathrm{i}}\right)-\dot{m}_{\mathrm{o}}\left(u_{\mathrm{o}}+\frac{v_{\mathrm{o}}^{2}}{2}+gz_{\mathrm{o}}\right)$$

where E_{cv} denotes the energy of the control volume at time t. The terms $\dot{Q}$ and $\dot{W}$ represent the net rate of energy transfer by heat and work across the boundary of the control volume at t respectively. $\dot{m}_{\mathrm{i}}u_{\mathrm{i}}$ is the internal energy of the entering stream, and $\dot{m}_{\mathrm{o}}u_{\mathrm{o}}$ is the internal energy of the exiting stream. $\dot{m}_{\mathrm{i}}\frac{v_{\mathrm{i}}^{2}}{2}$ is the kinetic energy of the entering stream, and $\dot{m}_{\mathrm{o}}\frac{v_{\mathrm{o}}^{2}}{2}$ is the kinetic energy of the exiting stream. $\dot{m}_{\mathrm{i}}gz_{\mathrm{i}}$ and $\dot{m}_{\mathrm{o}}gz_{\mathrm{o}}$ account for the potential energy of the entering stream and the exiting stream respectively.

2）as shown in Fig. …

例 1.2-13 For the one-dimensional flow as shown in Fig. 1.2-2, the energy rate balance can be written in the following form.

练习

1. 请使用 denote/represent/account for/mean/be 等动词各写一个英文句子。

2. 请用 as shown in Fig. … 写两个英文句子。

3. 请将下面的句子翻译为中文，并体会动词+ing 的用法。

1）The refrigerating effect of vaporization is more easily controlled with a vaporizing liquid.

2) The vaporizing temperature of liquid can be governed by controlling its vapor pressure.

3) Systems employing water-cooled condensers are divided into two categories.

4. 请将下面的句子翻译成中文，并体会情态动词+被动语态的用法。

1) The refrigeration machine can be started and shut down on demand.

2) The vapor can be condensed back into its liquid state.

5. 将下面的句子翻译为中文，并体会此处过去分词的用法。

1) Compressors convert the energy carried by electrical current into kinetic energy.

2) Water-cooled condenser employs water as the cooling medium.

1.3 Second Law of Thermodynamics

The first law of thermodynamics states that the cyclic integral of heat is equal to the cyclic integral of the work during any cycle that a system undergoes. Note that the first law places no restrictions on the direction of flow of heat and work, however, a cycle that does not violate the first law does not ensure that the cycle will actually occur. In fact, only the cycle that satisfies both the first and the second law of thermodynamics will occur.

There are two classical statements of the second law. They are the Kelvin-Planck and Clausius statements.

Here is the Kelvin-Planck statement. It is impossible to construct a device which operates in a cycle and produce no effect other than raising of a weight and exchange of heat with a single reservoir. It states that it is impossible to construct a heat engine that operates in a cycle, receives a given amount of heat from a high-temperature body, and does an equal amount of work. The only alternative is that some heat must be transferred from the working fluid at a lower temperature to a much lower body. Thus, work can be done by the transfer of heat only if there are two temperature levels, and heat is transferred from the high-temperature body to the heat engine and also from the heat engine to the low-temperature body. This implies that it is impossible to build a heat engine that has a thermal efficiency of 100%.

The Clausius statement is that it is impossible to construct a device that operates in a cycle and produces no effect other than transfer of heat from a cooler body to a hotter body.

This statement is related to the refrigerator or heat pump. It expresses that it is impossible to construct a refrigerator that operates without an input of work.

If it is impossible to have a heat engine of 100% efficiency, then what is the most efficient cycle that could be obtained?

The first step is to define a process that is called a reversible process. A reversible process for a system is defined as a process that can be reversed and leaves no change in either system or surroundings once it happened.

Many factors can make processes irreversible, such as friction, unrestrained expansion, heat transfer through a finite temperature difference, and mixing of two different substances.

The second step is to assume a cycle in which every process is reversible. Therefore, the cycle is reversible. The Carnot cycle is a reversible and the most efficient cycle that can operate between two constant-temperature reservoirs as shown in Fig. 1.3-1.

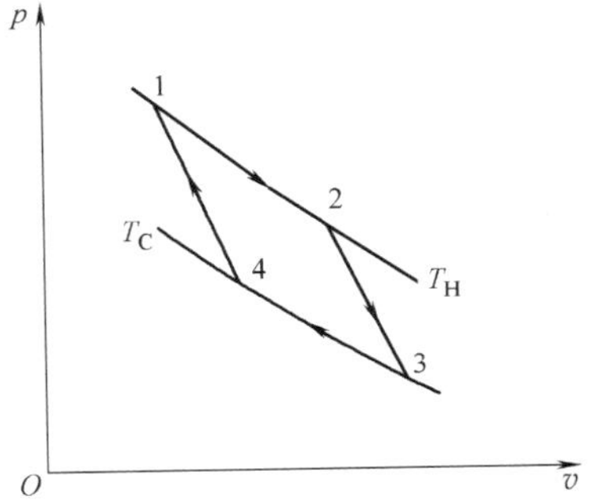

Fig. 1.3-1 *p-v* diagram for a Carnot cycle

If the cycle is a Carnot heat engine cycle, there are four processes in each cycle. The first process is a reversible isothermal expansion process in which heat is transferred from the high-temperature reservoir to the working fluid. The next process is an adiabatic expansion process during which the temperature of the working fluid decreases from the temperature of the high-temperature reservoir to the temperature of the low-temperature reservoir. In the third process, heat is rejected from the working fluid to the low-temperature reservoir. This must be a reversible isothermal process, in which the temperature of the working fluid is infinitesimally higher than that of the low-temperature reservoir. The final process, which completes the cycle, is a reversible adiabatic compression process in which the temperature of the working fluid increases from the low temperature to the high temperature.

Since Carnot cycle is reversible, every process could be reversed. If the heat engine cycle operates in reversed direction, the cycle becomes a refrigeration cycle. It should be noted that the efficiency of a real refrigeration cycle is always lower than that of the correspondent Carnot cycle due to unavoidable irreversible factors.

词汇表

second law of thermodynamics　热力学第二定律
cyclic　循环的，周期的
integral　积分
statement　表述
single reservoir　单一热源
heat engine　热机
imply　暗示
thermal efficiency　热效率
reversible　可逆的
reverse　逆行
irreversible　不可逆的
friction　摩擦，摩擦力
unrestrained expansion　自由膨胀
finite　有限的
temperature difference　温差
Carnot cycle　卡诺循环
isothermal　等温的
adiabatic　绝热的
infinitesimally　无限小地

unavoidable　不可避免的

学习要点

1. 宾语从句

科技论文中，常见到 that 引起的宾语从句（that 有时可省略）。

例 1.3-1　The first law of thermodynamics states that the cyclic integral of heat is equal to the cyclic integral of the work during any cycle.

这里 that 引导的从句 that the cyclic integral of heat is equal to the cyclic integral of the work during any cycle 作主句动词 state 的宾语，是宾语从句。

2. 定语从句

定语从句在科技论文中经常使用。

定语从句可分为限定性定语从句和非限定性定语从句两种。

1）限定性定语从句：用从句修饰前面的名词，从句是不能去掉的，去掉从句剩下的部分就不能清楚表达句子的意思。对一个专业词汇做定义时，常使用限定性定语从句。

例 1.3-2　Resistance temperature detector（RTD） is a common sensor which can measure temperature.

这里 which 引导的从句作定语修饰 a common sensor，which 在从句中作主语，不可省略。

例 1.3-3　In fact，only the cycle that satisfies both the first and the second law of thermodynamics will occur.

该例句中关系代词 that 引导的从句 that satisfies both the first and the second law of thermodynamics 就是限定性定语从句，修饰前面的名词 the cycle。这里的关系代词 that 在从句中作主语，指代 the cycle。这里 that 从句修饰的是“物”，这种情况 that 也可以用 which 代替。

2）非限定性定语从句：非限定性定语从句，是对前面修饰的名词的补充说明。非限定性从句去掉后，主句意思依然完整，不会受太大影响。在修饰“物”时用 which，不用 that，并且非限定性从句和被修饰词之间通常用逗号隔开。

例 1.3-4　This must be a reversible isothermal process，in which the temperature of the working fluid is infinitesimally higher than that of the low-temperature reservoir.

该例中，in which the temperature of the working fluid is infinitesimally higher than that of the low-temperature reservoir 是一个非限定性定语从句，是对前面的 a reversible isothermal process 的补充说明。非限定性定语从句和前面被修饰的名词间有一个逗号，并且这里用 which，不能用 that。

3. could be，must，possible 等词的准确应用

如果要叙述的事情有百分之百的肯定性可使用 must，completely，always，fully，thoroughly，totally，entirely，absolutely，definitely 等表示完全肯定的词，如果要叙述的

事情不完全肯定，只是可能的，要使用 possible，almost，could be 等弱肯定的词。在学术论文中准确使用这些表示不同肯定程度的词，可以帮助作者诚实准确地传递要表述的真实信息。

例 1.3-5 In the third process, heat is rejected from the working fluid to the low-temperature reservoir. This must be a reversible isothermal process, in which the temperature of the working fluid is infinitesimally higher than that of the low-temperature reservoir.

这里用到了 must，因为这个过程必须是可逆等温过程，否则就不是卡诺循环了，是百分之百的肯定。

例 1.3-6 A reversible process for a system is defined as a process that once it happened it can be reversed and leaves no change in either system or surroundings.

这里用了 can be reversed，没有用 must be reversed，表达可逆过程可以逆行，而不是必须逆行，用 can 比 must 更能准确表达作者要表达的意思，如果用得不准确就会导致歧义。

练习

1. 请下列句子翻译成英文。

1）卡诺循环的效率最高。

2）平衡蒸气压可以用实验测定。

2. 将下列英文句子翻译为汉语。

1）As a system undergoes a process, irreversibility may be found within the system as well as within its surroundings.

2）It is a matter of everyday experience that there is a definite direction for spontaneous processes. This can be seen by considering the three systems pictured in Fig. 1.

3. 下列哪些句子是限定性定语从句？哪些是非限定性定语从句？并说明句子中的that是否可以用which代替。

1） RED APPLE is a company that makes furniture.

2） Heat pump is a device that can supply both heating and cooling.

3） One choice to supply clean heating in winter is heat pump, which is driven by electricity.

4. 请使用that或which将以下两个句子组成一个复合句。

1） A building was destroyed in the fire. It has now been rebuilt.

2） A bus goes to the hospital. It runs every half hour.

3） The machine breaks down. It is now working again.

5. 请将下面的句子补充完整。

1） An electronic balance is an instrument that

____.

2) A thermocouple is a temperature sensor that

____.

3) A water pump is a device that

____.

4) An air conditioner is a device that

____.

5) A fan coil is a device that

____.

1.4 Heat Transfer

There are three modes of energy transfer. They are conduction, convection and radiation. All heat-transfer processes involve one or more of these modes.

Conduction is related with atomic and molecular activity. Conduction can be viewed as the transfer of energy from the more energetic to the less energetic particles of a substance due to interactions between the particles.

The basic equation to describe conduction is known as Fourier's law. For the one-dimensional plane wall shown in Fig. 1.4-1, the rate equation is expressed as

$$q''_x = -k\frac{\mathrm{d}T}{\mathrm{d}x} \tag{1.4-1}$$

where q''_x is the heat flux which is the heat transfer rate in the x direction per unit area perpendicular to the direction of transfer. Heat flux q''_x is proportional to the temperature gradient $\mathrm{d}T/\mathrm{d}x$. The parameter k is the thermal conductivity, which is a characteristic of the wall material.

Heat convection is the energy exchange between a surface and an adjacent fluid. Convections can be divided into forced and free convections. Forced convection is the flow which is driven by a fan or a pump. The flow that is caused by the density difference resulting from the temperature variation in fluid region is called free or natural convection.

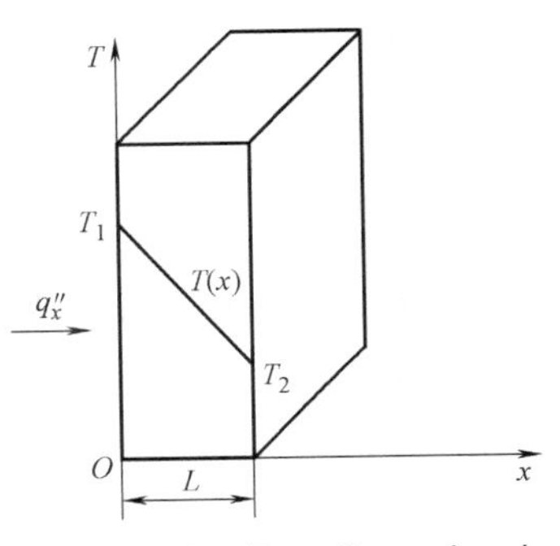

Fig. 1.4-1 One-dimensional heat transfer by conduction

The rate equation for convection is known as Newton's law of cooling. This equation is written as

$$\frac{q}{A} = h\Delta T \tag{1.4-2}$$

where q is the rate of convective heat transfer, in W; A is the area normal to direction of heat flow, in m^2; ΔT is the temperature difference between surface and fluid, in K; and h is the convective heat transfer coefficient, in $W/(m^2 \cdot K)$.

Radiation is the heat transfer between surfaces. It is different from conduction and convection. Note that no medium is required for its propagation.

The rate of radiation from radiator or black body is given by

$$\frac{q}{A} = \sigma T^4 \tag{1.4-3}$$

where q is the rate of radiation, in W; A is the area of emitting surface, in m^2; T is the absolute temperature, in K; and σ is the Stefan-Boltzmann constant, which is equal to 5.67×10^{-8} $W/(m^2 \cdot K^4)$.

词汇表

mode　方式
heat conduction　导热
convection　对流
radiation　辐射
atomic　原子的
molecular　分子的
energetic　有能量的
particle　粒子
Fourier's law　傅里叶定律
rate　率，比率，单位时间变化量
heat flux　热流密度
perpendicular　垂直的
proportional　成比例的
gradient　梯度
parameter　参数
thermal conductivity　热导率
characteristic　特性
forced convection　强制对流
free（natural）convection　自由对流
fluid　流体
fan　风机
pump　泵
normal to　垂直于
convective heat transfer coefficient　对流传热系数
propagation　传播
radiator　辐射体
black body　黑体
emit　发射
absolute temperature　绝对温度
Stefan-Boltzmann constant　斯特藩-玻尔兹曼常量

学习要点

1. 复合名词

在科技英语中，常见到复合名词，其通常由两个或多于两个名词组成。

例 1.4-1　heat transfer

该复合名词由两个名词组成，表示传热或热传递。

相似的用法本课文中还有 energy transfer, heat transfer coefficient, plane wall, rate equation, heat flux, heat transfer rate, heat convection, energy exchange, density difference, temperature variation, fluid region, temperature difference。

有时复合名词的几个名词之间用短画线相连，a three-day journey, a 300-seat auditorium, cottage-style kitchen window。

2. 动词的名词化 (nominalisation)

在科技英语中，为了使写作内容聚焦于概念和事物，常使用动词的名词化形式。因为采用动词是强调动作和事件，采用动词的名词化形式，是强调概念和事物。

例 1.4-2 The temperature distribution is measured with an infrared camera in this study.

这里 distribution 就是动词 distribute 的名词化形式。

3. 过去分词修饰名词

例 1.4-3 spoken and written English

4 现在分词修饰名词

例 1.4-4 an exciting moment

5. 公式及公式中各变量的说明

例 1.4-5 The rate of radiation from radiator or black body is given by

$$\frac{q}{A}=\sigma T^{4} \tag{1.4-3}$$

where q is the rate of radiation, in W; A is the area of emitting surface, in m^2; T is the absolute temperature, in K; and σ is the Stefan-Boltzmann constant, which is equal to 5.671×10^{-8} $W/(m^2 \cdot K^4)$.

练习

1. 英汉互译。

1)

$$\nu=\frac{\mu}{\rho}$$

这里 ν 是运动黏度，单位是 m^2/s；μ 是动力黏度，单位是 Pa · s；ρ 是密度，单位是 kg/m^3。

2）Under the steady-state conditions shown in Fig. 1. 4-1, the heat flux is

$$q''_x = -k\frac{T_2 - T_1}{L}$$

2. 阅读下面的一段英文并翻译为汉语。

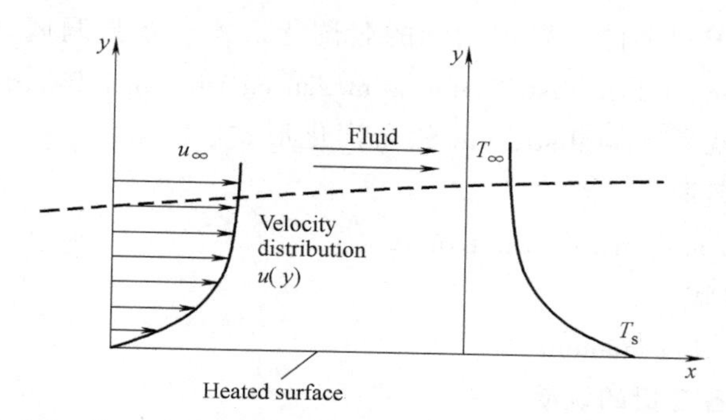

Fig. 1. 4-2 Boundary layer development in convection heat transfer

We are especially interested in convection heat transfer, which occurs between a fluid in motion and a bounding surface when the two are at different temperatures. Consider fluid flow over the heated surface of Figure 1. 4-2. A consequence of the fluid-surface interaction is the development of a region in the fluid through which the velocity varies from zero at the surface to a finite value u_∞ associate with the flow. This region of fluid is known as velocity boundary layer. Moreover, if the surface and flow temperature differ, there will be a region of fluid through which the temperature varies from T_S at $y=0$ to T_∞ in the outer flow. This region, called the thermal boundary layer, may be smaller, larger, or the same size as that through which the velocity varies [1].

3. 请写出下列动词的名词化形式。

consume

develop

select

separate

compare

transport

prepare
measure
collect
apply

4. 请写出几个复合名词。

5. 请写出几个过去分词修饰名词的例子。

6. 请写出几个现在分词修饰名词的例子。

1.5 Application of Mechanical Refrigeration

Refrigeration produces a temperature lower than the surroundings by removing heat through artificial methods. Cooling effects created by a mechanical device is mechanical refrigeration. The main application of mechanical refrigeration system is to preserve food. Another common application is to supply cold source for comfort or industrial air conditioning.

Application of mechanical refrigeration can be roughly classified into five categories: domestic refrigeration, commercial refrigeration, industrial refrigeration, marine and transportation refrigeration, and air conditioning. Overlapping parts may exist between different categories because the boundaries of these categories are not set precisely. Some application can be put in more than one category.

Domestic refrigeration includes all household refrigerators and freezers that are basically insulated boxes. Cooling effects are normally realized by vapor compression refrigeration systems. Fig. 1.5-1 shows a typical household refrigerator. A typical freezer is shown in Fig. 1.5-2.

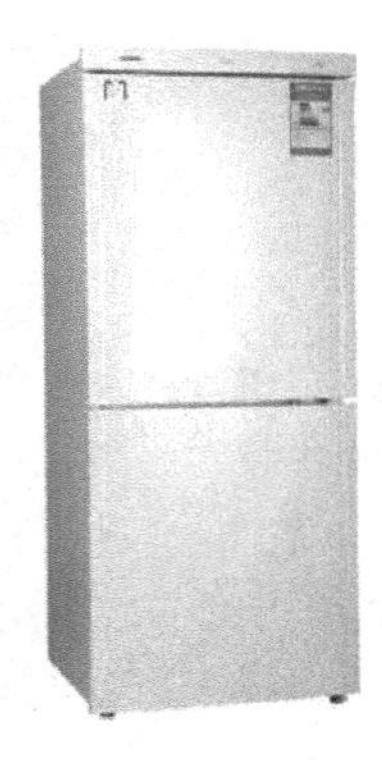

Fig. 1.5-1 Household refrigerator

Fig. 1.5-2 Freezer

The cooling capacity and size of these devices are usually small. An induction motor is generally hermetically sealed within a steel compressor dome.

Commercial refrigeration is widely used in retail stores, supermarkets, restaurants, hotels, laboratories, hospitals etc. These units are used to store, display and process perishable things such as food, medicine etc. Large reach-in refrigerators and walk-in coolers are typical commercial refrigeration equipment. Fig. 1.5-3 is a typical reach-in refrigerator. Reach-in refrigerators can be easily found in supermarkets. Some reach-in refrigerators can be used not only for storage but also display purposes. A walk-in cooler as shown in Fig. 1.5-4 allow two- or four-wheel carts to move freely through the door and load large boxes.

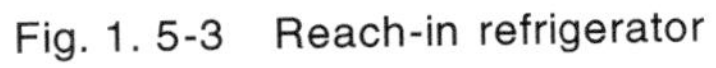

Fig. 1. 5-3 Reach-in refrigerator

Fig. 1. 5-4 Walk-in cooler

Industrial refrigeration applications are often confused with those in commercial refrigeration because the concept of these two areas is not clearly defined. Normally, industrial devices are larger than the commercial ones. Operating engineers are always needed to be on duty to be responsible for the safe operating. Typical industrial applications can be found in ice-making plants, cold stores, breweries and creameries. Fig. 1. 5-5 shows the typical industrial refrigeration systems used for large scale cold stores.

Fig. 1. 5-5 Typical industrial refrigeration systems

Refrigerating and freezing systems used in fishing boats, vessels, trucks or refrigerated railway cars that ship perishable cargo fall into the category of marine and transportation refrigeration. Note that there are also comfort air conditioning systems on boats, trains, trucks, airplanes as well. They belong to the category of air conditioning. Fig. 1. 5-6 is a typical refrigerated truck.

Air conditioning system controls air temperature, humidity, air velocity and air cleanliness automatically. Air conditioning applications are classified as either comfort or industrial ones according to the design intention. Comfort air conditioning is a system designed to satisfy human need for comfort. They are commonly found in homes, hotels, supermarkets, large department stores, cinemas, gymnasiums, hospitals, automobiles, buses, trains, planes and ships. Industrial

air conditioning systems can be used in textile factories, data centers and printing houses etc. The aims of industrial air-conditioning systems can be summarized as follows:

1) To improve the quality of products.

2) To limit the size variations due to thermal expansion and contraction.

3) To provide trouble-free operating environment for equipment and instruments.

4) To create specific experimental conditions.

Fig. 1. 5-6 A typical refrigerated truck

词汇表

refrigeration 制冷
application 应用
cold source 冷源
comfort air conditioning 舒适空调
industrial air conditioning 工艺空调
domestic refrigeration 家用制冷
commercial refrigeration 商用制冷
industrial refrigeration 工业制冷
marine and transportation refrigeration 船舶和运输制冷
category 范畴
boundary 边界
set 设置
refrigerator 冰箱，冷柜
freezer 冰柜
basically 根本上
insulate 隔热
vapor compression refrigeration system 蒸气压缩式制冷系统
cooling capacity 制冷量
induction motor 感应电动机
hermetically 密封地
seal 密封
steel 钢制的
compressor 压缩机
dome 圆顶
process 加工

perishable 易腐的
reach-in refrigerator 陈列式冷柜
walk-in cooler 步入式冷藏库
operating engineer 运行工程师
ice-making plant 制冰厂
cold store 冷库
brewery 酿酒厂，啤酒厂
creamery 乳品厂
humidity 湿度
gymnasium 体育馆
thermal expansion 热膨胀
contraction 收缩
textile factory 纺织厂
data center 数据中心
summarize 概述

学习要点

1. 过去分词短语作定语

本课文中的几个句子都使用了过去分词短语作定语。

例 1.5-1 Cooling effects created by a mechanical device is mechanical refrigeration.

该例句中 created by a mechanical device 是过去分词短语，用来修饰前面的 cooling effect。该句也可用定语从句的形式表达，写为：

Cooling effects that are created by a mechanical device is mechanical refrigeration.

但用过去分词的形式表达更简洁。

2. 动词不定式作表语

科技英语中动词不定式作表语较常见。

例 1.5-2 One aim of industrial air-conditioning systems is to improve the quality of products.

这里不定式短语 to improve the quality of products 在句子中作表语。

3. 冒号的使用

冒号一般较少使用，但在下面的场合，可以使用冒号。比如归纳或总结出几条，罗列这几条之前，通常使用冒号。

例 1.5-3 The aims of industrial air conditioning systems can be summarized as follows:

1) To improve the quality of products.

2) To limit the size variations due to thermal expansion and contraction.

3) To provide trouble-free operating environment for equipment and instruments.

4) To create specific experimental conditions.

练习

1. 将下面的句子翻译为英语。

1）制冷的主要用途是保存食物。

2）为提高产品质量，纺织厂通常都安装集中空调系统（central air conditioning systems）以调节空气的温度、湿度和清洁度。这种空调系统被称为工艺空调。

3）机械制冷的应用可划分为家用、商用、工业用、运输制冷和空调。

4）制冷就是将物体温度降低和维持到环境温度以下。

5）冷水机组（water chiller）制备的冷冻水（chilled water）给集中空调系统提供冷源。

2. 将下面的英文句子翻译为汉语。

1） Walk-in coolers are main food storage fixtures.

2） The principal function of any kind of display fixture is to exhibit the product or commodity as attractively as possible in order to stimulate sales.

3. 写出下列词汇的同义词或近义词。

device　classify　include　observe

4. 写出下列动词的名词化形式。

apply　classify　describe　vary　expand　contract

5. 请找出课文中不定式短语作表语的句子。

1.6 Vapor Compression Refrigeration

The vapor compression refrigeration is the most widely used refrigeration method. The application can be found in domestic and commercial refrigerators, large-scale cold storage warehouses for chilled or frozen storage of foods, air conditioners, water chillers and refrigerated trucks. The evaporator, compressor, throttling device and condenser are the four primary components of a vapor compression system as shown in Fig. 1. 6-1. The components are connected with copper or steel tubes, and the pipe joints must be perfectly sealed. The refrigeration loop must be evacuated before charging refrigerant. A certain amount refrigerant must be charged before it operates.

A refrigeration vapor compression cycle is made up of four fundamental processes: expansion, vaporization, compression and condensation. The refrigerant experiences state changes in its pressure, temperature and phase during each cycle.

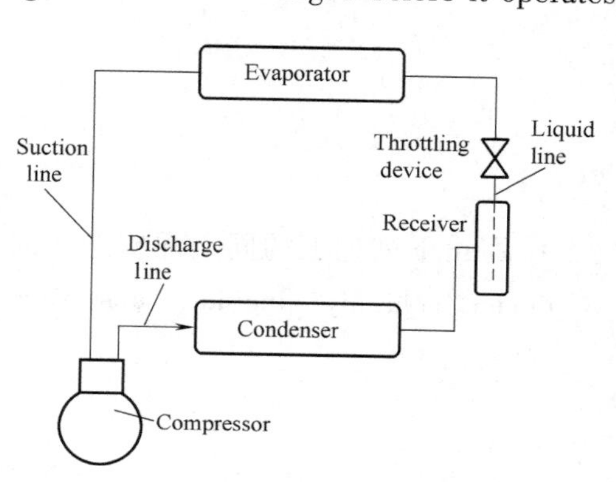

Fig. 1. 6-1 A vapor compression refrigeration system

The liquid refrigerant at high pressure and high temperature in the receiver flows into the liquid line and then enters the inlet of the throttling device. As the refrigerant flows through the throttling device, the pressure of the refrigerant decreases from condensation to evaporation pressure. This throttling process causes some of refrigerant to flash. Consequently, the temperature of the remaining liquid decreases. After flowing through the throttling device, the mixture of liquid and flash vapor enters the evaporator. Fig. 1. 6-2 shows three typical throttling devices including thermostatic expansion valve, electronic expansion valve and capillary tube.

Fig. 1. 6-2 Throttling devices

The refrigerant vaporizes in evaporator, which is a low-pressure heat exchanger. The evaporation of refrigerant absorbs the heat from the cooled space. In this process, the cooling effect occurs. The vapor from the evaporator is normally superheated to avoid liquid entering the compressor.

Compression of the refrigerant occurs in compressor. The refrigerant vapor is sucked into the compressor because a low pressure zone is generated as the compressor runs. A compressor consists of an electric motor and a vapor compression apparatus. During the compression process, the temperature and pressure of the vapor increase. The high-temperature and high-pressure vapor is discharged into the discharge line and enters the condenser.

A condenser is a high-pressure heat exchanger. Air-cooled and water-cooled condensers are the most common types. The refrigerant vapor condenses in condenser as the temperature of cooling air or water is lower than the refrigerant vapor's saturation temperature. The heat of condensation is released in this process. When water is used as a cooling medium, cooling towers are often necessary to cool the heated water from the condenser. Cooling water flows through the condenser, carrying the released heat of vapor condensation and enters the cooling tower. The heated water is sprayed and vaporized in the cooling tower so that the remaining water is cooled and driven back to the condenser by a water pump. A slightly subcooled refrigerant leaves the condenser, flows back to the receiver and returns to its initial state to complete a cycle. The cycle repeats and repeats to supply a continuous cooling.

The vapor compression refrigeration cycle can be described and analyzed easily using a pressure-enthalpy diagram (p-h diagram). A basic saturated vapor compression cycle is a theoretical cycle with following assumptions:

1) The outlet of evaporator and inlet of compressor are saturated vapor.

2) The outlet of condenser or receiver and inlet of throttling device are saturated liquid.

The theoretical cycle can be drawn on a p-h diagram in Fig. 1.6-3. Point 1 to 2 is a compression process. Point 2 to 3 is a condensation process. Point 3 to 4 is a throttling or an expansion process. Point 4 to 1 is an evaporation process. The cooling capacity per unit mass refrigerant Q can be calculated with Eq. (1.6-1) where h_1 is specific enthalpy of point 1 and h_4 is specific enthalpy of point 4. The specific work for a compression process W can be calculated with Eq. (1.6-2) where h_2 is specific enthalpy of point 2. The coefficient of performance COP of the cycle can be calculated with Eq. (1.6-3).

$$Q = h_1 - h_4 \quad (1.6\text{-}1)$$

$$W = h_2 - h_1 \quad (1.6\text{-}2)$$

$$COP = Q/W = (h_1 - h_4)/(h_2 - h_1) \quad (1.6\text{-}3)$$

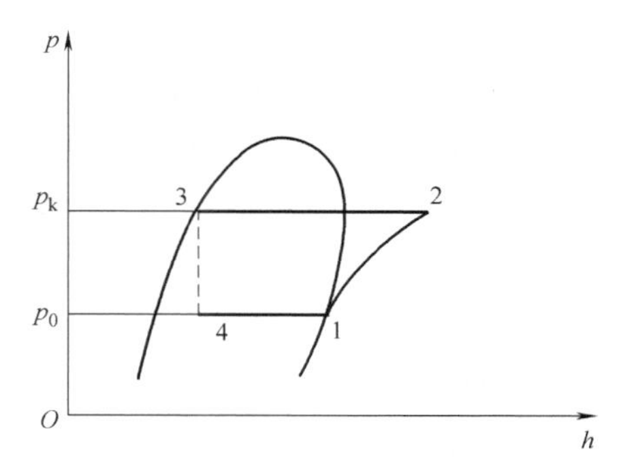

Fig. 1.6-3 A theoretical cycle

Based on Fig. 1.6-3, three ways are listed here to achieve a higher COP:

1) As evaporation pressure p_0 increases, the cooling capacity Q increases and the specific work for compression W decreases, so COP increases.

2) As condensation pressure p_k decreases, the cooling capacity Q remain unchanged and W decreases, so

COP increases.

3) As evaporation pressure p_0 increases and condensation pressure p_k decreases, as a result, Q increases, W decreases and COP increases.

Therefore, for an air conditioned room, setting the room temperature to 28℃ can effectively save energy instead of setting it to 18℃.

The actual cycle normally deviates from these saturated conditions. The cycle with liquid subcooling is shown in Fig. 1. 6-4. If evaporation and condensation pressures keep constant, and the specific work is same, comparison between Fig. 1. 6-3 and Fig. 1. 6-4 shows cooling capacity per unit mass of the latter is greater than the former. Therefore, the COP will increase if the subcooled liquid enters the throttling device instead of saturated liquid.

Fig. 1. 6-5 shows the cycle with vapor superheating. When the superheating of vapor has cooling effect, it is called effective superheating. Cooling capacity Q_0 equals to ($h_{1'}-h_4$), which is greater than (h_1-h_4) of a theoretical cycle. If the superheating has no cooling effect, the cooling capacity Q_0 is still (h_1-h_4). And specific work W is ($h_{2'}-h_{1'}$) for both. The increase or decrease of COP is dependent on the property of refrigerant. If the refrigerant is carefully selected, the effective superheating cycle can provide better COP.

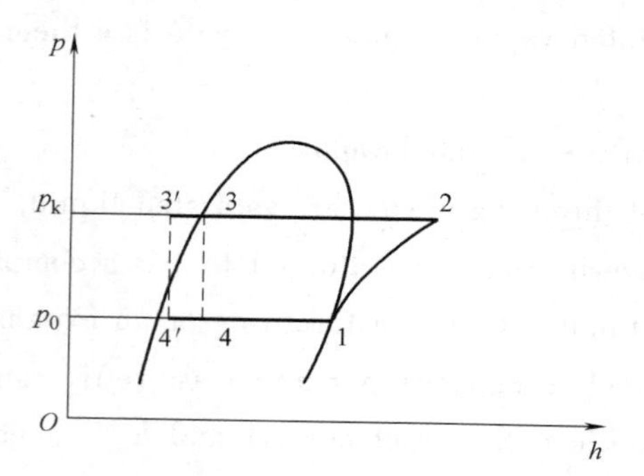

Fig. 1. 6-4 Cycle with liquid subcooling

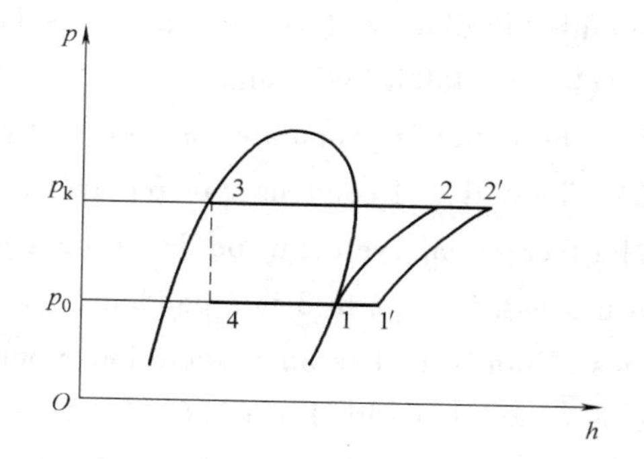

Fig. 1. 6-5 Cycle with vapor superheating

The cycle with a recuperator is shown in Fig. 1. 6-6 and Fig. 1. 6-7. The high temperature liquid from condenser and low temperature vapor from evaporator exchange heat in a recuperator. The result is the temperature of suction vapor increases and inlet temperature of throttling valve decreases. COP can be calculated with Eq. (1. 6-4). The increase or decrease of COP is dependent on the property of refrigerant. If the refrigerant is suitably selected, better COP will be reached.

$$COP = Q/W = (h_1 - h_{4'})/(h_{2'} - h_{1'}) \qquad (1.6\text{-}4)$$

All above-mentioned cycles are called single-stage vapor compression cycles. When evaporation temperature is too low, the compression ratio is bigger than 8-10. The refrigerant vapor is difficult to be sucked into compressor due to influence of clearance volume. In this situation, two-stage, multistage and cascade vapor compression cycles can be found in application.

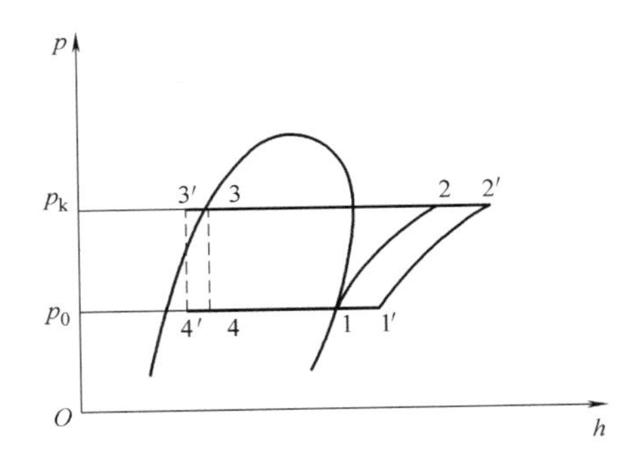

Fig. 1. 6-6　Cycle with a recuperator

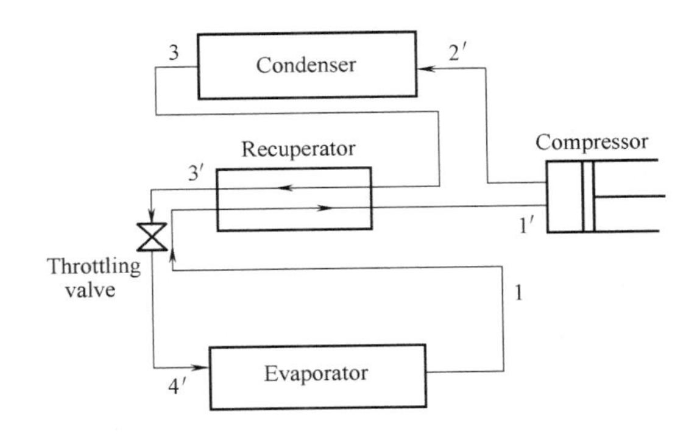

Fig. 1. 6-7　Schematic of a cycle with a recuperator

词汇表

large-scale cold storage warehouses　大型冷库
chilled food　冷藏食品
frozen food　冷冻食品
air conditioner　空调器
water chiller　冷水机组
refrigerated truck　冷藏车
evaporator　蒸发器
condenser　冷凝器
throttling device　节流装置
component　部件
pipe joint　管接头
refrigeration loop　制冷回路
refrigerant　制冷剂
charge　充注
expansion　膨胀
vaporization　蒸发
compression　压缩
condensation　冷凝
liquid line　液体管线
saturation　饱和
flash　闪发
thermostatic expansion valve（TXV）　热力膨胀阀
electronic expansion valve　电子膨胀阀
capillary tube　毛细管
suck　吸

electric motor　电动机
apparatus　设备
discharge line　排气管线
receiver　储液器
heat exchanger　热交换器
air-cooled condenser　空冷冷凝器
water-cooled condenser　水冷冷凝器
cooling tower　冷却塔
spray　喷雾
vaporize　蒸发
water pump　水泵
subcooled　过冷的
initial state　初始状态
inlet　入口
outlet　出口
pressure-enthalpy diagram　压焓图
assumption　假设
per unit mass refrigerant　单位质量制冷剂
calculate　计算
specific work　比功
evaporation pressure　蒸发压力
actual cycle　实际循环
deviate　偏离
saturated liquid　饱和液体
superheating　过热
property　性质
recuperator　回热器、间壁式换热器
single-stage vapor compression cycle　单级蒸气压缩循环
clearance volume　余隙容积
two-stage vapor compression cycle　两级蒸气压缩循环
multistage vapor compression cycle　多级蒸气压缩循环
cascade vapor compression cycle　复叠蒸气压缩循环

学习要点

1. 被动语态

语态分为主动语态和被动语态。在科技英语中经常使用被动语态，有时还使用带情态动词的被动语态。

例 1.6-1　I performed these experiments.

该例句是主动语态。但科技英语中需要客观表达要描述的事物，要尽量避免使用人称代词，所以通常使用被动语态。该例句可改写为被动语态如下：

These experiments were performed.

科技英语中为了准确表达描述的事物，还常使用带情态动词的被动语态。

例 1.6-2　The refrigeration loop must be evacuated before charging refrigerant.

这里强调充注制冷剂前制冷剂回路必须抽真空。为强调必须，这里使用了 must。

例 1.6-3　The application can be found in domestic and commercial refrigerators, large-scale cold storage warehouses for chilled or frozen storage of foods, air conditioners, water chillers, refrigerated trucks.

该例句中使用了情态动词 can，而不是 must。强调可能性，因此用 can 而不用 must。

2. 使用图表和文字说明某个过程

为了使表达更清楚，本课文中大量使用了图表，并参照图表写说明文字，请注意学习。

练习

1. 将下列句子翻译为英语。

1）蒸气压缩式制冷系统由四个主要部件组成，即压缩机、蒸发器、冷凝器和节流装置。

2）蒸气压缩式制冷系统的工作过程常用压焓图描述。

3）冷凝器有水冷冷凝器和空冷冷凝器两种类型。

4）冷却塔主要用来给水冷冷凝器的冷却水降温。

5）制冷剂在蒸发器内低压下蒸发以吸收周围空间的热量。

6）蒸气压缩式制冷系统包括下列类型：单级、双级、多级和复叠式制冷系统等。

2. 请用英语描述图 1.6-4 上过冷循环的工作过程。

3. 请找出课文中出现被动语态的句子，并说明为什么这里要用被动语态。

4. 请用 can 或 must 填空。

1） The experimental result ________ be explained with Brownian movement.

2） The uncertainty analysis ________ be given in research paper for experimental study.

3） The whole experimental system ________ be rebuilt to satisfy the basic requirement of future experiment.

4） In this situation, two-stage, multistage and cascade vapor compression cycles ______ be found in application.

1.7 Heat Pump

Heat pump is one type of refrigeration device. It not only produces cooling but also heating effects. Heat pumps can be based on either vapor compression refrigeration or absorption refrigeration. To design a vapor compression heat pump, HVAC&R engineers must consider the following points:

1) The indoor coil must have a larger surface area compared with that of a single-cold air conditioner to prevent the condensation temperature from becoming too high in heating mode.

2) The air-handling capacity of indoor unit and duct system must allow sufficient air flow to assure adequate condensation of the refrigerant.

3) The compressor must be different from a single-cold air conditioning compressor and must be specially designed for heat pump application because it operates all year long at completely different operating pressures and temperature conditions.

4) Almost all heat pump systems use a suction line accumulator to prevent liquid refrigerant from entering the compressor crankcase during the off mode. Scroll compressors are more suitable to be used in heat pumps.

5) Defrost control is required to keep the unit operating at high efficiency.

6) Auxiliary heating elements must be designed to support the unit work in extreme cold days.

There are two main modes for vapor compression heat pump operation, namely cooling mode and heating mode. In the cooling mode, the indoor coil works as an evaporator and the outdoor coil works as a condenser. In the heating mode, the reversing valve changes position to reverse the refrigerant flow. The heat is absorbed into the system by vaporizing the refrigerant in outdoor coil. Then a low-temperature and low-pressure vapor flows to the compressor. In the compressor, the refrigerant is compressed to high-pressure and high-temperature vapor, and is discharged to the indoor coil. The energy that is equal to the value of the absorbed heat from outdoor coil plus compression work is released through indoor coil to the conditioned space.

Vapor compression heat pumps can be classified into four types: water-water, water-air, ground-air and air-air. The first part of these combinations is referred to as the heat source of the outdoor coil. The second part refers to the medium treated by the refrigerant in the indoor coil.

Water-water heat pump uses water to heat or cool the outdoor coil and the required warm or chilled water is treated by the indoor coil. Sources of water for the outdoor coil can be wastewater, or water from a single well, double wells, a lake or a pond. Water-air heat pump uses similar water source like water-water ones for outdoor coil and air is treated by the indoor coil. The only difference of the above-mentioned two types is the indoor section. The former produces a certain temperature of water to be used and the latter maintains the air at a desired temperature. For a ground-air heat pump, the outdoor coil is buried underground. The heat is extracted from the soil. In most seasons the temperature of the soil remains about 18℃ at about 1.8m below the ground level.

The disadvantages of this system are as follows:

1) If refrigerant leaks, the soil around the coils will be polluted.

2) The ground tends to be loose around the coils due to expansion and contraction of the coils.

Air-air heat pump is the most popular type. They are easier and more economical to be installed, and maintenance costs are also less than water-water or water-air units. However, air-air systems consume more electricity than water-source heat pumps.

In winter, especially for an air-source heat pump, the outside air transfers heat to the outdoor coil and further to the refrigerant. Even though the outside air temperature maybe above 0℃, the moisture in the air will freeze and form frost on the surface of the outdoor coil, since the surface temperature of the outdoor coil may be decreased below the freezing temperature of water. The frost on the coil can restrict the air from passing through the coil. The frost will also act as an insulator on the finned surface and reduce heat transfer. Consequently, the system performance is significantly affected. Therefore, defrost is very important for heat pump. Frost can be removed in the cooling mode by changing switch position of the reversing valve. Simultaneously the outdoor fan stops running. The hot refrigerant vapor discharged from compressor is directed to the outdoor coil to melt the frost. There are several methods of automatic defrost. Control variables are listed as follows:

1) Air pressure differential across the outdoor coil

2) Outdoor coil temperature

3) Time

4) Time and temperature

Proper defrost control can improve the efficiency of the heat pump unit effectively.

词汇表

heat pump 热泵

manufacturer 生产企业

air conditioning unit 空调机组

valve 阀门

reverse 逆转

efficiency 效率

HVAC&R (heating, ventilating, air conditioning and refrigeration) 暖通空调和制冷

single-cold air conditioning 单冷空调

heating mode 加热模式

air-handling capacity 空气处理能力

indoor unit　室内机组
duct system　风道系统
accumulator　气液分离器
crankcase　曲轴箱
reciprocating compressor　往复式压缩机
scroll compressor　涡旋压缩机
liquid slugging　液击
defrost　除霜
auxiliary　辅助的
heat source　热源
indoor coil　室内盘管
outdoor coil　室外盘管
indicate　表示
medium　介质
treat　处理
chilled water　冷冻水
above-mentioned　上述的
former　前者
latter　后者
extract　抽取
namely　也就是
reversing valve　换向阀
vaporize　蒸发，汽化
moisture　水气，湿气
insulator　绝热体，绝缘体
finned surface　翅片表面
consequently　结果
pressure differential　压差

学习要点

1. 长句句子分析

科技论文中经常遇到长句。只有分辨出主句、从句，理解分词及动词不定式的用法，才能很好理解长句表达的意思。

例 1.7-1　The indoor coil must have a larger surface area compared with that of a single-cold air conditioner to prevent the condensation temperature from becoming too high in heating mode.

该例句中出现的动词有四个：have，compare，prevent，become。但只有一个主句，

The indoor coil must have a lager surface area to prevent the condensation temperature from becoming too high in heating mode，过去分词短语 compared with that of a single-cold air conditioner 修饰前面的 surface area。不定式短语 to prevent the condensation temperature from becoming too high in heating mode 在主句中作状语表示目的。

例 1.7-2　The energy that is equal to the value of the absorbed heat from outdoor coil plus compression work is released through indoor coil to the conditioned space.

该例句的主句为 The energy is released through indoor coil to the conditioned space，从句 that is equal to the value of the absorbed heat from outdoor coil plus compression work 作定语修饰前面的名词 energy。

2. prevent ... from doing sth.

例 1.7-3　The indoor coil must have a larger surface area compared with that of a single-cold air conditioner to prevent the condensation temperature from becoming too high in heating mode.

这里 from 后面要跟动词的 ing 形式，这是固定用法。

例 1.7-4　Almost all heat pump systems use a suction line accumulator to prevent liquid refrigerant from entering the compressor crankcase during the off mode.

这里 from 后面跟 enter 的 ing 形式，这是固定用法。

3. 归纳与总结的写法

常用写法如下，注意学习。

Control variables are listed as follows:

1）Air pressure differential across the outdoor coil

2）Outdoor coil temperature

3）Time

4）Time and temperature

练习

1. 将下面的句子翻译为英语。

1）在冬季不太寒冷的地区，热泵可代替暖气供暖。

2）夏季可用空调模式降低室温，冬季可用热泵模式升高室温。

3）热泵的除霜是制冷空调领域的一个研究热点。

4）热泵可分为四种类型：水-水热泵、水-空气热泵、地源-空气热泵、空气-空气热泵。

5）好的控制策略可帮助热泵节能。

2. 请用英语口头描述一下热泵的组成、工作原理及应用。用学术英语撰写出相应的内容，并写出日常英语和学术英语的不同。

3. 阅读并翻译下面的短文。

ACCUMULATOR OR RECEIVER?

The tank on the suction line between the evaporator and the compressor is a suction accumulator. The tank on the liquid line between the condenser and TXV is a liquid receiver. They do look similar but they serve two completely different purposes.

The primary function of the suction accumulator is to catch and hold any liquid refrigerant that didn't boil off in the evaporator. Liquid refrigerant getting to the compressor can damage the pistons or scrolls. This liquid will also dilute or even flush the oil out of the compressor crankcase. This loss of oil will prevent proper lubrication to the compressor, causing compressor damage or failure. Liquid slugging can occur even on a properly installed system with the loss of gas flow. Improper evaporator gas flow due to dirty filters, coil or loose belt will have the same effect. Low suction temperatures such as on a heat pump in the heating mode can also cause liquid slugging of the compressor. Many heat pump manufactures utilize suction accumulators as standard equipment.

The accumulator function is quite simple. The suction gas leaving the evaporator enters the accumulator at the top and passes through a baffle or screen. Any liquid present collects on the screen and falls to the bottom of the accumulator. Inside the accumulator is a U shaped tube that will allow only the refrigerant vapor to exit and enter the compressor. A small orifice in the bottom of the U tube will allow any oil that collected in the accumulator to exit and return to the compressor through the suction line. Accumulator failures are rare on properly maintained systems. A plugged orifice in the U tube would be the most likely problem. This plugged orifice would prevent oil from returning to the compressor.

An accumulator is inexpensive and can be added to almost any system that has experienced compressor slugging. The cause of the slugging should still be determined and corrected if possible. Systems that run under low load conditions may be a good place to add an accumulator. Parker recommends that the accumulator is replaced when a compressor is being replaced. Contaminated oil from the old compressor may be in the old accumulator. Also, a considerable amount of oil may still be in the old accumulator. This oil combined with the oil from the new compressor may create an oil overcharge.

Proper accumulator sizing is important when replacing or adding. The pressure drop across the accumulator should be kept as low as possible. The accumulator's internal volume must be sufficient. On a heat pump system with a fixed metering device the accumulator should be capable of holding 70% of the system charge. In a TXV system, the accumulator should be able to handle 50% of the system charge. Refer to sizing charts for proper sizing. The accumulator should never be sized by connection sizes [2].

1.8 Evaporators and Condensers

Heat exchanger is a device whose primary function is to transfer energy between two fluids. Evaporators and condensers are typical heat exchangers in refrigeration systems. Evaporators can be manufactured in different types, shapes and sizes to meet a variety of applications. Evaporators for air-cooling are called air coolers, while evaporators for water-cooling are called water chillers.

Air coolers can be classified into bare tube, plate-surface or finned-tube type based on the construction as shown in Fig. 1. 8-1.

They can also be divided into dry-expansion, flooded and circulated refrigerant type based on the method of refrigerant feed as shown in Fig. 1. 8-2.

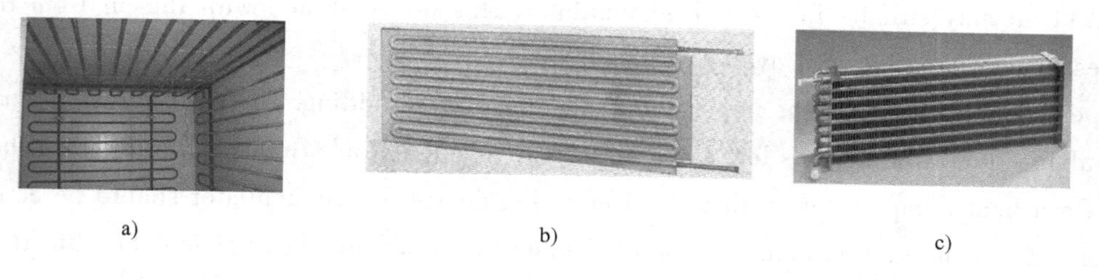

a) b) c)

Fig. 1. 8-1 Types of evaporators

a) Bare tube b) Plate-surface c) Finned-tube

Dry-expansion evaporators are normally used in residential and commercial refrigeration systems, while flooded and circulated refrigerant types are normally employed in large commercial and industrial applications. For the dry-expansion type, the amount of refrigerant fed into the evaporator is limited to the quantity that can be completely vaporized when the refrigerant enters the suction line. The metering device employed with this feed method is either a thermostatic expansion valve (TXV) or a capillary tube. Circulated refrigerant evaporators incorporate a liquid refrigerant pump to circulate liquid refrigerant through the evaporator circuits. The pump draws liquid refrigerant from the bottom of an accumulator. A flooded evaporator is similar to a circulated refrigerant evaporator. They both have an accumulator and are nearly filled with liquid refrigerant during the operation. The difference locates in method of refrigerant circulation, the accumulator position and the diameter of the pipe used in evaporator. The accumulator of a flooded evaporator is positioned so that the level maintained by the float is at the same elevation as the centerline of the top coil row. This allows the evaporator to remain flooded while still allowing the vapor to exit the coil through the upper half of the top tube. However, the elevation of the accumulator is not critical for circulated refrigerant evaporators because a liquid pump is employed in this kind of system.

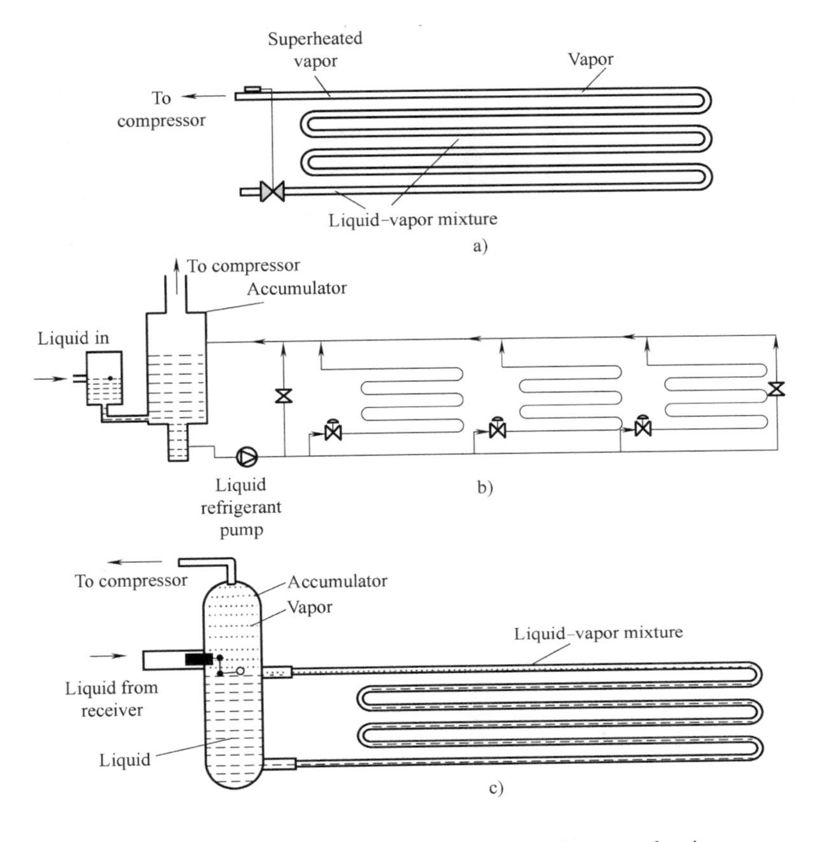

Fig. 1. 8-2 Evaporators classified by refrigerant feed

a) Dry-expansion evaporator b) Circulated refrigerant evaporator c) Flooded evaporator

Condensers transfer heat from the refrigerant to the cooling medium. Superheated refrigerant vapor can be changed into subcooled liquid by removing the heat with the cooling medium. There are three types of condensers: air-cooled, water-cooled and evaporative-cooled. Air-cooled condenser employs air as the cooling medium to absorb heat from the refrigerant. The circulation of air may be by natural or forced convection. Water-cooled condenser employs water as the cooling medium. There are three types of water-cooled condensers, namely tube-in-tube, shell-and-coil and shell-and-tube as shown in Fig. 1. 8-3. The tube-in-tube or double-tube condenser consists of two tubes arranged in the form that one is suspended within the other. Cooling water flows through the inner tube while the refrigerant flows in the annular space between the inner and outer tubes. The shell-and-coil condenser is made up of one or more bare-tube or finned-tube coils enclosed within a welded steel shell. The cooling water circulates through the coils while the refrigerant is contained within the shell. The shell-and-tube condenser is a cylindrical steel shell in which a number of straight tubes are arranged and held in place by tube sheets. The cooling water is circulated through steel or copper tubes and refrigerant is contained within the steel shell.

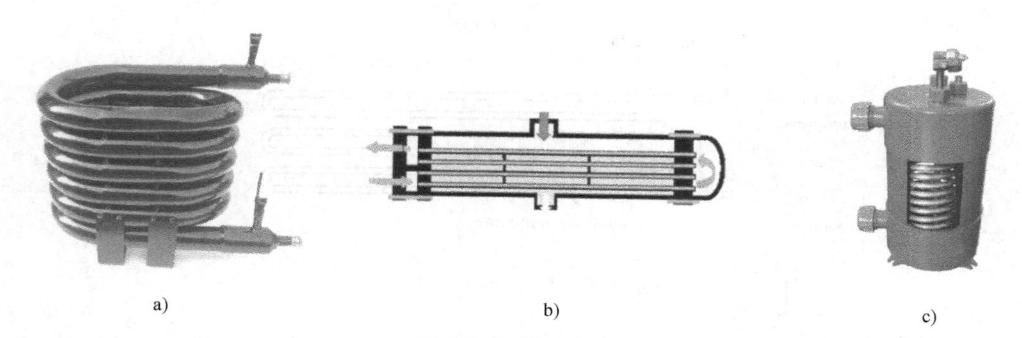

Fig. 1. 8-3 Types of water-cooled condensers

a) Tube in tube b) Shell-and-tube c) Shell-and-coil

词汇表

device 设备
function 功能
air cooler 空气冷却器，冷风机
bare tube 光管
plate-surface type 板式
finned-tube type 翅片管式
construction 结构
dry-expansion 干膨胀
flooded evaporator 满液式蒸发器
circulated refrigerant type 循环制冷剂型
refrigerant feed 制冷剂供液
metering device 节流装置
incorporate 包括
float 浮子，浮球
elevation 高度
cooling medium 冷却介质
superheated 过热的
evaporative-cooled type 蒸发冷却型
natural convection 自然对流
tube-in-tube type 套管型
shell-and-coil 壳盘管型
shell-and-tube 壳管型
cylindrical 圆筒形的
straight tube 直管
tube sheet 管板

学习要点

1. whose、who、whom 及 where 引出的定语从句

除了有 that、which 引出的定语从句，还有 whose、who、whom 及 where 引出的定语从句。

例 1.8-1 Heat exchanger is a device whose primary function is to transfer energy between two fluids.

该例句中 whose primary function is to transfer energy between two fluid 是 whose 引出的从句，用来修饰前面的名词 device。whose 表示……的，这里表示设备的。

例 1.8-2 A scientist who works at University of Chicago proposes this new concept.

该例句中 who works at University of Chicago 是 who 引出的定语从句，说明前面的 scientist 是什么样的科学家，是在芝加哥大学工作的科学家。这里 scientist 是人，所以 who、whom 或 whose 可能出现在从句中。这里的关系代词在从句中作主语，所以从句使用 who。

例 1.8-3 The professor whom I wanted to see is waiting for me at his office.

该例句中 whom I wanted to see 是 whom 引导的定语从句，修饰 professor。在该定语从句中 whom 作 see 的宾语，因此这里使用 whom。

例 1.8-4 The shell-and-tube condenser is a cylindrical steel shell in which a number of straight tubes are arranged and held in place by tube sheets.

该例句中 in which a number of straight tubes are arranged and held in place by tube sheets 是定语从句，修饰前面的 a cylindrical steel shell，这里是说明 a cylindrical steel shell 里面有什么，或者说 a cylindrical steel shell 那个地方有什么，在这种情况，in which 可用 where 代替。

2. 复杂长句分析

例 1.8-5 For the dry-expansion type, the amount of refrigerant fed into the evaporator is limited to the quantity that can be completely vaporized when the refrigerant enters the suction line.

该例句中主句为 the amount of refrigerant is limited to the quantity, fed into the evaporator 是过去分词短语作定语修饰前面的 the amount of refrigerant, that can be completely vaporized 是 that 引导的定语从句修饰前面的 the quantity, when the refrigerant enters the suction line 是时间状语从句。该例句既有主句又有从句，因此为复合句。

3. 被动语态

学术论文中经常使用被动语态，本文中也有多处使用。因为学术论文一般要客观陈述事实，一般动作的主体是谁并不重要。通常学术论文中要避免使用 I、we、you 等人称代词。

练习

1. 请将下列语句改为被动语态，并体会在学术论文中应该用哪种语态。

1） We performed a sequence of experiments.

2） We built an experimental setup as shown in Fig. 1.

2. 请将下面一段英文翻译为汉语。

While refrigerant boils inside the tubes of most commercial evaporators, in one important class of liquid-chilling evaporator the refrigerant boils outside the tubes. This type of evaporator is standard in centrifugal compressor applications. Sometimes such an evaporator is used in conjunction with reciprocating compressors, but in such applications provision must be made for returning oil to the compressor. In the evaporators where refrigerant boils in the tubes, the velocity of the refrigerant vapor is maintained high enough to carry oil back to the compressor ［3］.

3. 请用 that/which/who/whom/whose/where 填空。

1） A teacher is someone ____________ teaches courses.

2） I don't like movies ____________ have unhappy endings.

3） The building ____________ was destroyed by earthquake has now been rebuilt.

4） I found the keys ____________ you lost.

5） A cinema is a place ____________ people watch movies.

6） That is the man ____________ car was towed away by the police.

7） The woman with ____________ I work is very nice.

8） I went back to the small town ____________ I grew up last month.

1.9 Air Conditioning

Air conditioning is the process of altering the state of air to specified conditions. Namely, the air temperature, humidity, cleanliness and flow velocity are regulated to satisfy the need of human comport or production process. The former is comfort air conditioning and the latter is industrial air conditioning.

Mechanical ventilation has close relation with air conditioning systems. The air conditioning system differs from ventilation by the incorporation of refrigeration system. Therefore, it can be said that the mechanical ventilation turns into air conditioning by adding mechanical refrigeration equipment and cooling coils.

Air conditioning can be used in following situations:

1) High room temperature in summer.

2) Workshop is not clean enough for manufacturing electronics.

3) The temperature and humidity in the workshop affects the quality of products significantly such as paper and textile production.

4) Some buildings have to be air-sealed from the external environment to limit noise penetration, consequently require mechanical ventilation and necessary refrigeration.

5) Data centers must be air conditioned to remove the large heat released from operating servers.

6) Shops, hotels and other commercial buildings etc. where comfort environments are attractive to people.

7) Storage and display of paintings, antique, furniture, fabrics and paper archives etc.

8) Sterile conditions required for medical treatment, such as operating rooms in hospital.

Air conditioning systems can be divided into three categories based on cooling medium:

1) Total air systems

2) Air-water systems

3) Refrigerant systems

Total air system regulates the room temperature with handled air. This system is also called central air conditioning system. The air is handled by mixing, filtering, cooling, heating etc. in an air handling unit. Then the handled supply air is distributed into different air conditioning rooms by air ducts.

Air-water system regulates the space temperature with water and air together. Fan-coil plus fresh air system is the typical air-water system which is normally used in guest rooms in hotels. Cooling load is settled mainly by the flowing chilled water in the cooling coil. The fresh air can be handled and supplied to the conditioned space by air duct to satisfy the requirements of human health.

Residential air conditioner is the typical refrigerant system. The cooling load in the conditioned space is treated by evaporation of refrigerant. No machine room is needed to handle the air. It is convenient to install split and packaged units in homes or offices etc. Therefore, refrigerant systems are widely used and can be seen easily in hot summer area.

Air is the object to be handled in air conditioning engineering. To analyze state change of air easily, the concept humid air is defined in the study of air conditioning to represent mixture of dry air and moisture. Dry air in atmosphere includes 78% nitrogen, 21% oxygen by volume and a tiny amount of other gases such as carbon dioxide and inert gases etc. Any state of humid air can be found a point on a psychrometric chart. The air handling process can be expressed clearly by a psychrometric chart. There are two kinds of commonly used psychrometric charts in the world. One was proposed by Willis Carrier, who also invented the first modern electrical air conditioning unit. This kind of psychrometric chart is used in the USA and Europe. Another is proposed by Richard Mollier. Mollier's psychrometric chart is used in China, Russia etc. These two kinds of psychrometric chart are very similar. Parameters like dry-bulb temperature, wet-bulb temperature, relative humidity, moisture content, specific enthalpy, dew point temperature can be easily found on a psychrometric chart.

The outside temperature varies all the time. Normally the daytime temperature is higher than the night-time temperature. The temperature inside the building is normally different from outside. The heat gain or loss is determined mainly by the followings:

1) Heat transfer through glazing
2) Heat transfer through exterior walls and roofs
3) Heat transfer through ceilings, floors and interior walls
4) Heat generated in the space by people, lights and devices
5) Outdoor fresh air and exhaust air cause heat gain or loss

The cooling load is the quantity of heat that must be removed from the conditioned space to maintain the desired room temperature and humidity. The cooling load is not heat gain. Heat gain refers to the quantity of heat that is transferred into or generated inside a building. Heat gain includes sensible and latent heat. The latent heat and heat convection can change into cooling load immediately. The radiation heat can't change into cooling load immediately. It becomes cooling load through a decay and delay process. The radiation heat is normally absorbed by the interior wall first and causes the temperature of interior wall to rise. When the temperature of interior wall is higher than the surroundings, the heat is released to the room by heat convection and turns into cooling load. Cooling load calculation is very important to design air conditioning systems.

To save equipment and operating costs, it is very important to select the suitable type of air conditioning systems according to the actual situation of the construction site. Equipment selections are also very important for energy saving. To avoid selecting oversized refrigerating devices, cooling load calculation is essential for suitable equipment selection.

Single-duct variable air temperature with recirculation system is used commonly in high-rise office buildings. The refrigeration devices and air handling units are normally placed in a machine room as shown in Fig. 1. 9.

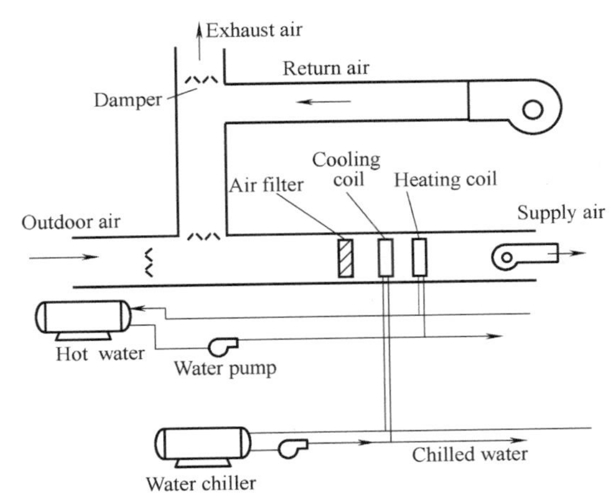

Fig. 1. 9 Devices in an air conditioning machine room

The vapor compression or absorption water chillers are used as cold source. Boilers can be used as heat source. An air handling unit is a main device to deal with the air. Firstly, the outdoor air and return air is mixed. Secondly, mixed air is filtered. Thirdly, clean air is cooled to apparatus dew point (ADP). Fourthly, the air is reheated to required temperature of supply air. Then the supply air is distributed to different rooms by supply-air fan through air ducts. The return air is driven by a return air fan to return-air duct. A small amount of return air, which is equal to the amount of outdoor fresh air, is emitted to atmosphere to maintain the indoor air pressure. The rest goes back to the inlet of the system and mix with outdoor fresh air. This process repeats and repeats to supply handled air continuously. When the cooling load changes with time, the water flow rate of hot water or chilled water must also be regulated automatically to adapt the change of cooling load. So automatic control systems are also very important for air conditioning systems.

词汇表

cleanliness 清洁度
velocity 速度
regulate 调节
ventilation 通风
incorporation 合并
textile 纺织品
air-sealed 空气密闭的

penetration 穿入，渗入
server 服务器
access 进入
antique 古董
fabrics 纤维织物
sterile 无菌的
medical treatment 医疗
operating room 手术室
category 范畴
total air system 全空气系统
central air conditioning system 集中空调系统
air duct 风道
fan coil 风机盘管
cooling load 冷负荷
residential air conditioner 家用空调
machine room 机房
split and packaged units 分体和一体式空调机
humid air 湿空气
dry air 干空气
nitrogen 氮气
oxygen 氧气
carbon dioxide 二氧化碳
psychrometric chart 焓湿图
propose 提出
dry-bulb temperature 干球温度
wet-bulb temperature 湿球温度
relative humidity 相对湿度
moisture content 含湿量
apparatus dew point 机器露点
heat gain 得热量
loss 损失
determine 确定
exterior wall 外墙
roof 屋顶
interior wall 内墙
lights 灯，灯光，照明
fresh air 新风

exhaust air　排风
sensible heat　显热
latent heat　潜热
decay　衰减
delay　延迟
equipment cost　设备费用
operating cost　运行费用
construction site　施工工地
single-duct variable air temperature with recirculation system　单风道变空气温度回风系统
boiler　锅炉
supply air　送风

学习要点

描述过程，使用 firstly、secondly、thirdly 等表示时间顺序的副词。

例 1.9　Firstly, the outdoor air and return air is mixed. Secondly, mixed air is filtered. Thirdly, clean air is cooled to apparatus dew point (ADP). Fourthly, the air is reheated to required temperature of supply air. Then the supply air is distributed to different rooms by supply air fan through air ducts. The return air is driven by a return air fan to return-air duct.

练习

1. 请参照图 1.9 用英文叙述一次回风的集中空调系统的工作过程，并使用 firstly、secondly、thirdly 等表示时间顺序的副词。

2. 请将下面一段话翻译成英文。

这个小旅馆总冷负荷只有 20kW，空调可采用风机盘管系统。风机盘管的冷热源可采用风冷热泵机组。也可在各房间安装空调器。但集中空调系统不适合用在这种场合。

1.10 Absorption Refrigeration

Absorption refrigeration is a heat driven refrigeration technology. The waste heat from chemistry factories and power plants can be used to drive absorption chillers to supply required chilled or hot water. The solar energy can also be used for absorption refrigeration. It is an energy saving and environment friendly refrigeration mode. Especially electrical energy consumption can be reduced greatly.

An absorption cycle is similar to the vapor compression cycle because both use a volatile refrigerant that alternately vaporizes under low pressure in the evaporator and condenses under high pressure in the condenser. The main difference is the methods employed to circulate the refrigerant through the system and maintain the pressure difference between the evaporator and condenser. Compressor is required in a vapor compression system, but in an absorption system it is not. Instead of compressor, a solution circulation system is used to increase the pressure of refrigerant vapor.

A simple diagram of a single-effect absorption cycle is depicted in Fig. 1.10. There are four basic components. On the low-pressure side, there are an evaporator and an absorber. On the high-pressure side, there are a generator and a condenser. In addition to the above-mentioned four components, a solution heat exchanger is also an important component for energy saving purpose.

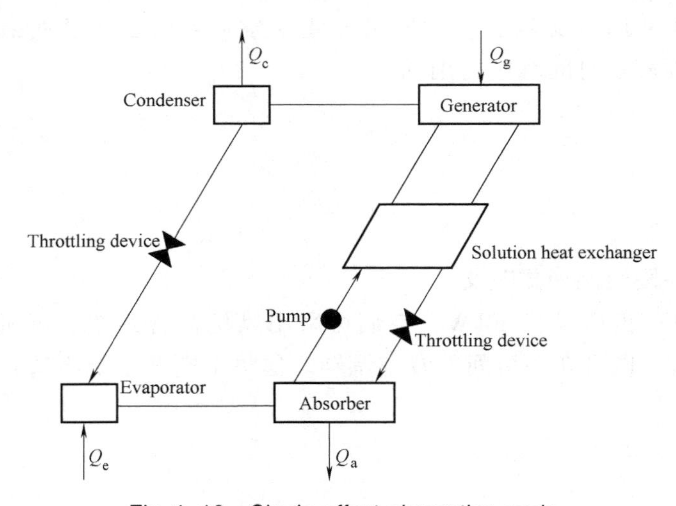

Fig. 1.10 Single-effect absorption cycle

The absorption working fluid is called working pair which includes a refrigerant and an absorbent. The most commonly used working pairs are ammonia-water and water-lithium bromide. Ammonia-water absorption systems are widely used in commercial and industrial chillers where the evaporation temperature can be lower than 0℃. These chillers use ammonia as refrigerant and water as absorbent. Water-lithium bromide absorption chillers are used extensively as cold source for central air conditioning systems in buildings. In these chillers, water works as refrigerant and lithium bromide as absorbent.

The absorption systems can be direct-fired, steam, flue gas, hot water types based on the heat source type. There are two on-going cycles when an absorption system is operating. One is a refrigerant cycle and the other is a solution cycle. For a water-lithium bromide absorption system, the refrigerant cycle works as follows. The high temperature refrigerant vapor from generator flows into condenser where sensible and latent heat is removed by cooling water, and vapor is changed into liquid. After leaving condenser, liquid pressure is reduced by a throttling device. The low pressure refrigerant enters evaporator where it is changed into vapor and absorbs heat from the refrigerated space. Then the refrigerant vapor is absorbed in absorber by solution from generator.

After absorption, the poor solution in absorber is pumped to the generator. The refrigerant vapor is generated as the poor solution is heated by heat source in generator and flows to the condenser to complete a refrigerant cycle.

Before the solution cycle is stated, a set of terms must be taken good care of. The terms "rich" and "poor" are sometimes used but care must be taken to know which component these terms refer to. Similar terms are "strong" and "weak". For a water-lithium bromide solution, the poor solution normally means poor in absorbent.

To complete the solution cycle, the high temperature and high pressure rich solution from the generator exchanges heat with the low temperature poor solution from absorber in solution heat exchanger. The low temperature rich solution passes a throttling device to reduce pressure to low pressure and enters absorber to absorb the refrigerant vapor. The generated poor solution is pumped to the generator.

During absorption process, the absorption heat is released. The released heat must be removed efficiently to keep the temperature, pressure and absorption rate in absorber nearly constant. The absorption rate influences unit COP seriously. Scientists have realized that intensifying heat and mass transfer in absorber is the key point to increase the unit COP. Lots of experiments have been performed for increasing absorption rate and heat transfer by adding surfactants or developing high efficiency heat exchange tube etc.

Falling film absorbers are commonly used in large absorption chillers. A solution distributor is fixed on the top of the absorber. The parallel tube bundles are placed just under the distributor. The cooling water flows in the tube to remove the absorption heat. The rich solution from generator falls down from the top of absorber through a liquid distributor. During falling, the refrigerant vapor is absorbed and the temperature of solution increases. The solution touches the outside surface of cooling tube during falling down. The absorption heat is removed effectively by this way.

The surfactant such as n-octanol or 2E1H can be added in the solution to enhance absorption and heat transfer by inducing Marangoni convection. After leaving the absorber, the cooling water flows through the condenser and carry the condensation and absorption heat to cooling tower. In cooling tower, the heat is transferred to the ambient.

Absorption cycle can be divided into single-effect, double-effect and triple-effect cycle depending on the temperature of heat source in generator by times of vapor generation. If the temperature of heat source is about 100℃, the solution can generate vapor only once and this kind of cycle is called single-effect cycle. If the temperature of heat source is around 150℃, the two-effect cycle can be designed to increase COP. The cooling COP is about 0. 7 for a single-effect water-lithium bromide system and 1. 0-1. 2 for a double effect water-lithium bromide system.

词汇表

absorption refrigeration　吸收式制冷
power plant　电厂
solar energy　太阳能
energy saving　节能
environment　环境
volatile　易挥发的
alternately　交替地
circulate　使循环
single-effect absorption cycle　单效吸收循环
depict　描述，描绘
component　部件
solution heat exchanger　溶液热交换器
working fluid　工质
absorbent　吸收剂
working pairs　工质对
lithium bromide　溴化锂
ammonia　氨
generator　发生器
absorber　吸收器
direct-fired absorption chiller　直燃式吸收机组
flue gas　烟气
steam　水蒸气
poor solution　稀溶液
rich solution　浓溶液
term　术语，词语
absorption rate　吸收率

intensify 加强，强化
key point 关键点
perform experiments 做试验
surfactant 表面活性剂
heat exchange tube 换热管
develop 开发
falling film absorber 降膜吸收器
solution distributor 溶液布液器
fix 使固定，安装
parallel 平行的
fall 落下
n-octanol 正辛醇
2E1H 异辛醇
Marangoni convection 马兰戈尼对流
ambient 周围环境
double-effect 双效
triple-effect 三效

学习要点

1. 过去分词的使用

例 1.10-1 The main difference is the methods employed to circulate the refrigerant through the system and maintain the pressure difference between the evaporator and condenser.

该例句中 employed 引出的过去分词短语，用来修饰前面的名词 methods。该句也可改为下面的定语从句表达，定语从句中使用了被动语态。但用过去分词表达更简洁。

The main difference is the methods that are employed to circulate the refrigerant through the system and maintain the pressure difference between the evaporator and condenser.

2. 动词不定式 to+动词和介词 to

例 1.10-2 The low temperature rich solution passes a throttling device to reduce pressure to low pressure and enters absorber to absorb the refrigerant vapor.

该例句中第一次出现 to，to reduce pressure 是动词不定式短语，作目的状语，to 后面是动词 reduce。第二次出现的 to 是介词，to 后面是名词。第三次出现 to，to absorb the refrigerant vapor 是动词不定式短语，作目的状语，to 后面是动词 absorb。

3. 表示位置的介词 in/at/on

例 1.10-3 There are many people in the small coffee shop. It is very crowded.

Go straight along this road, then turn right at the coffee shop.

可以看出第一句中介词 in 是在……内，而第三句中 at 是在……地点

例 1.10-4 There is a map on the wall.

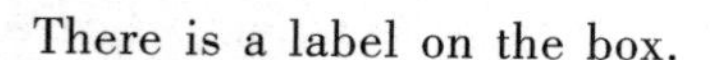

There is a label on the box.

There are some books in the box.

可以看出 on 表示在……表面上，in 表示在……内。

4. 介词 to 的用法

介词 to 有多种用法，下面仅介绍常用的几种。

例 1.10-5 以例 1.10-2 为例。

该例句中 to low pressure 中的 to 表示到（某时间、某数量或某状态）。

例 1.10-6 Let's go to school.

这里的 to 表示向（某处）、去（某地）。

例 1.10-7 You are so nice to us.

这里的 to 表示对……，to 后面是人称代词。

练习

1. 阅读下面的短文，并翻译为汉语。

Water/lithium bromide is an absorption working fluid which has been used widely since 1950s when the technology was pioneered by several manufacturers in the U. S. This working fluid utilizes water as the refrigerant and is therefore limited to refrigeration temperatures above 0℃. Absorption machines based on water/lithium bromide are typically configured as water chillers for air conditioning systems in large buildings. Machines are available in sizes ranging from 10 to 1500 Ton（Note：Ton is a unit of refrigeration capacity，1Ton = 12000Btu/h = 3. 517kW）. The coefficient of performance（COP）of these machines，defined as the refrigeration capacity divided by the driving heat input，typically varies over the range 0. 7<COP<1. 2 depending on the particular cycle configuration. These machines have a reputation for consistent，dependable service among mechanical room operators. The main competing technology is vapor compression chillers and choice between the two depends strongly on economic factors［4］.

2. 将下列句子翻译为英文。

1）溴化锂-水吸收系统存在的主要问题是溶液的腐蚀性（corrosion）和结晶问题（crystallization）。

2）氨水吸收系统的主要问题是必须使用精馏器（rectifier）以分离制冷剂氨蒸气中的水。

3）吸收器是吸收式系统中最重要的部件。

3. 请用动词的适当形式填空。

1）Five thermocouples are evenly fixed in the outer wall of the test channel ______ (measure) the temperature change of the fluid along the channel.

2）These five thermocouples ______ (fix) in the outer wall of the test channel are copper constantan thermocouples.

4. 请改正下列句子的语法错误。

1）The woman lives next door is a teacher.

2）The experiment was performed use an infrared camera.

3）They will go attend a party this evening together.

4）She is suitable being a mechanical engineer.

5. 请用介词 in/at/on 填空。

1) The hotel is ______ an island.

2) The book is ______ the table.

3) There is a notice ______ the door.

4) Do you know the girl standing ______ the window?

5) You can check in ______ reception.

6) He lives ______ Beijing.

7) A glass bottle filled with water ______ the thermostatic waterbath is a vapor generator.

6. 请说明下面句子中的 to 是介词还是动词不定式。

1) He will go to Qingdao next month.

2) The sealed tube system is filled with high pressure nitrogen to check for gas leaks.

3) Experiments are also performed to verify the results of numerical study.

4) The required temperature range is from 5℃ to 10℃.

5) It is very difficult to keep the pressure in the test section at a certain value.

1.11 Refrigerants

Refrigerant is the working fluid that circulates in the refrigeration system. Cooling effect is produced by vaporizing refrigerant in evaporator.

At early time, diethyl ether, ammonia, carbon dioxide, sulfur dioxide are all used as refrigerants. But diethyl ether is explosive and dangerous. Ammonia is a colorless gas with a pungent smell. It is toxic, flammable and explosive. Ammonia is usually used in large cooling capacity machines due to its low price and excellent thermodynamic property. Carbon dioxide can also be used in ships due to its non-toxicity. The machines used carbon dioxide as refrigerant are usually very heavy.

In the early 1930s, Thomas Midgley invented the chlorofluorocarbon (CFC) R12. This refrigerant and other member of CFC family are very suitable to be used as refrigerants. They are stable, nontoxic, non-flammable, with good thermodynamic properties and oil miscibility. The CFCs R12, R11, R114 and R502 together with hydrochlorofluorocarbon (HCFC) R22 became the most frequently used refrigerants. These CFCs and HCFCs are normally called freon, which is a registered trademark of the Chemours Company.

In 1974, scientists found the mechanism of depletion of ozone layer. Afterwards, it is found CFCs and HCFCs are ozone depletion substances and greenhouse gases. It is the chlorine in CFCs and HCFCs that causes the chain reaction with ozone and thins the ozone layer. Halons and bromides are also ozone depletion substances.

The ozone depletion potential (ODP) of a refrigerant represents its effect on atmospheric ozone. The ODP for R11 is 1, and this is the reference point adopted for all refrigerants. For example, the ODP for R12 is 0.82, this means that the effect of R12 on ozone layer is 0.82 times of effect of R11.

After a series of meetings and negotiations, the Montreal Protocol on Substances that Deplete the Ozone Layer was finally agreed in 1987. Signatories agreed to phase out the production of several CFCs and halons by 1995. The refrigeration industry rapidly moved from CFCs to HCFCs. At subsequent revisions of the Protocol, a phase-out schedule for HCFCs was also set. R22, which is an HCFC, has a far lower ODP than the CFCs but is a greenhouse gas. Under the Protocol, HCFCs will be eliminated by 2030.

To replace the chlorine in CFCs and HCFCs, the chemical companies developed a range of hydrofluorocarbons (HFCs). R134a, which is the first HFC to become available, is a close match to R12. The other HFC refrigerants now in wide use are blends of two or three HFCs.

The released greenhouse gas into atmosphere causes global warming, which is possibly the most severe environmental issue today. Global warming can cause extreme weather, sea level rise and species extinction. The most infamous greenhouse gas is carbon dioxide, which remains in the atmosphere for 500 years once released. A major CO_2 emission comes from fossil fuel power generation.

The global warming potential (GWP) of a gas is defined as an index comparing the climate impact of its emission to that of emitting same amount of carbon dioxide. For example, R134a has a GWP of 1300, this means that 1kg emission of R134a is equivalent to 1300kg of CO_2.

The Kyoto Protocol is the most famous international treaty to reduce greenhouse gas emissions. It was adopted in Kyoto Japan on December 11, 1997 and entered into force on February 16, 2005.

The natural refrigerants with zero ODP and low GWP represent a long-term solution to the environmental issue of refrigerant leakage. Apart from natural refrigerants, mixed refrigerants are also used to replace freon. R407C, R410A and R404A are most frequently used non-azeotropic mixture.

R407C consists of 23% R32, 25% R125 and 52% R134a. It is close to properties of R22, therefore it has been extensively used for replacing R22.

R410A is the mixture of 50% R32 and 50% R125. Its theoretical performance is poor. However, the refrigerant side heat transfer is better than that with R22. The pressure drop in equivalent heat exchangers is less. Research has shown that the COP of systems optimized for R410A is better than R22 equivalent systems.

R404A has been designed for commercial refrigeration where it is now widely applied. It has superior performance to the other HFCs in low temperature applications.

Hydrocarbons such as propane and butane have also been successfully used in systems where CFCs and HCFCs have previously been employed. They have obvious flammable characteristics that must be taken into account. However, there is a large market for their use in sealed refrigerant systems such as domestic refrigeration and air conditioners.

Nowadays, carbon dioxide as refrigerant attracts much interest again. High latent heat and heat transfer coefficient combined with high pressure and density under operating conditions results in the ability to produce large amounts of cooling with very small displacement compressors and small diameter pipelines. The transcritical cycle is very effective which can be used in carbon dioxide heat pump water heater. Carbon dioxide can also be used in vehicle air conditioning systems.

词汇表

diethyl ether　乙醚

sulfur dioxide　二氧化硫

explosive　有爆炸性的

pungent　刺鼻的

toxic　有毒的

flammable 易燃的
invent 发明
chlorofluorocarbon（CFC） 氯氟烃
Midgley 米基利（氟利昂的发明者）
stable 稳定的
miscibility 可混合性，相溶性
hydrochlorofluorocarbon（HCFC） 含氢氯氟烃
freon 氟利昂
mechanism 机理
ozone layer 臭氧层
depletion 损耗
chlorine 氯
halon 哈龙
bromide 溴化物
ozone depletion potential 臭氧消耗潜能值
reference point 参考点
adopt 采纳
a series of 一系列的
negotiation 谈判
Montreal Protocol 蒙特利尔议定书
signatory 签署者，签署国
phase out 使逐步淘汰，逐渐停止
subsequent 随后的
revision 修订，修正案
hydrofluorocarbons（HFCs） 氢氟烃
blend 混合物
global warming 全球变暖
environmental issue 环境问题
greenhouse gas 温室气体
infamous 臭名昭著的
fossil fuel 化石燃料
power generation 发电厂
global warming potential（GWP） 全球变暖潜值
index 指数
Kyoto Protocol 京都议定书
natural refrigerant 天然制冷剂
leakage 泄漏

non-azeotropic　非共沸的
hydrocarbon　碳氢化合物
propane　丙烷
butane　丁烷
pressure drop　压降
superior　优秀的，出众的
displacement compressor　容积式压缩机
transcritical cycle　跨临界循环
vehicle air conditioning　车辆空调

学习要点

1. 学习课文中使用的限定性定语从句和非限定性定语从句
2. 介词 for 的用法

例 1.11-1　R407C consists of 23% R32, 25% R125 and 52% R134a. It is close to properties of R22, therefore it has been extensively used for replacing R22.

该例句中 for 是为了（表示目的）的意思。介词 for 后是动名词短语 replacing R22。

例 1.11-2　for example

for example 为固定搭配，表示举例。

例 1.11-3　This is for you.

这里的 for 是给……的。

3. 介词 by 的用法

例 1.11-4　Cooling effect is produced by vaporizing refrigerant in evaporator.

介词 by 后面跟一个动名词短语，by 表示通过某种方式。

例 1.11-5　Under the protocol, HCFCs will be eliminated by 2030.

这里介词 by 后跟年份，表示到（某时候）。

例 1.11-6　He goes to work by bus.

这里 by 表示乘车方式。

例 1.11-7　He is bitten by a dog.

被动语态的句子中，by 表示被……的意思，by 后跟动作的主体。

4. 介词 with 的用法

例 1.11-8　Ammonia is a colorless gas with a pungent smell.

介词 with 后面是名词，这里 with 是带有、具有……的意思。

例 1.11-9　The CFCs R12, R11, R114 and R502 together with hydrochlorofluorocarbon (HCFC) R22 became the most frequently used refrigerants.

这里的 together with，表示和……在一起的意思，with 后面是名词。

例 1.11-10　It is the chlorine in CFCs and HCFCs that causes the chain reaction with ozone and thins the ozone layer.

这里的 with 是和、与的意思，with 后面是名词。

例 1.11-11　Its theoretical performance is poor. However, the refrigerant side heat transfer is better than that with R22.

这里的 with 是用的意思。

例 1.11-12　With the developing of technology, CO_2 can be used as a wonderful natural refrigerant in refrigeration devices.

这里的 with 是随着的意思。

练习

1. 请将下列汉语句子译成英语。

1）二氧化碳常用于热泵热水器。

2）大型冷库经常使用氨为制冷剂。

3）水是一种天然制冷剂，可用于蒸气压缩式制冷系统中。

4）溴化锂-水吸收式制冷系统中，水是制冷剂，溴化锂是吸收剂。

2. 请将下列英语翻译为汉语。

1）Low GWP and ODP natural refrigerants such as water and CO_2 are welcome nowadays.

2）With the development of technology, CO_2 can be used as a wonderful natural refrigerant

in refrigeration devices.

3. 请用 by/with/for 填空。

1) Of these refrigerants, less than a dozen are often encountered _____ a service technician.

2) Methane series refrigerants were developed _____ replacing some or all of the hydrogen atoms _____ chlorine or fluorine atoms.

3) Methylene chloride was developed _____ use in centrifugal compressors.

1.12 Refrigeration Compressors

A refrigeration compressor provides a driven force for a vapor compression system. Compressors can be divided into hermetic, semi-hermetic and open types. The drive motor and compressor of an open compressor are two separate components. A driveshaft protrudes from the side of the compressor's crankcase through a seal. The seal not only prevents refrigerant and oil from escaping, but also keeps moisture and air from entering the system. Open compressors are usually used in large cooling capacity applications. All ammonia refrigerant systems use open compressors. Fig. 1. 12-1 is an open compressor manufactured by GEA.

An electric drive motor and a compressor sealed within a steel shell is called a hermetic compressor, which is not accessible for repair or maintenance. The compressor is coupled with the motor's shaft. All small and some medium cooling capacity refrigeration compressors are hermetic. Fig. 1. 12-2 shows typical hermetic compressors.

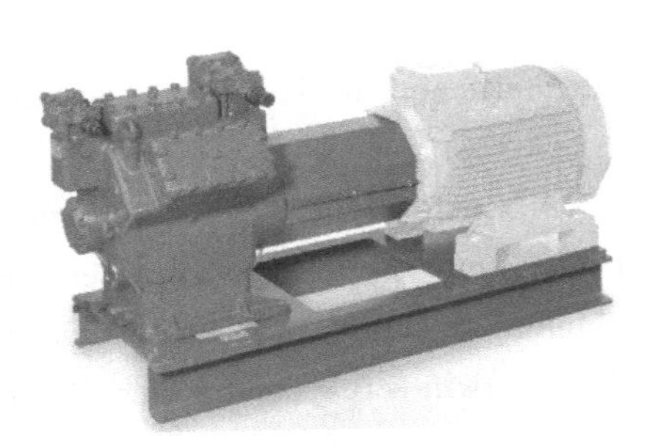

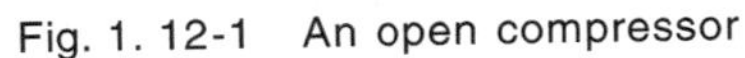

Fig. 1. 12-1 An open compressor

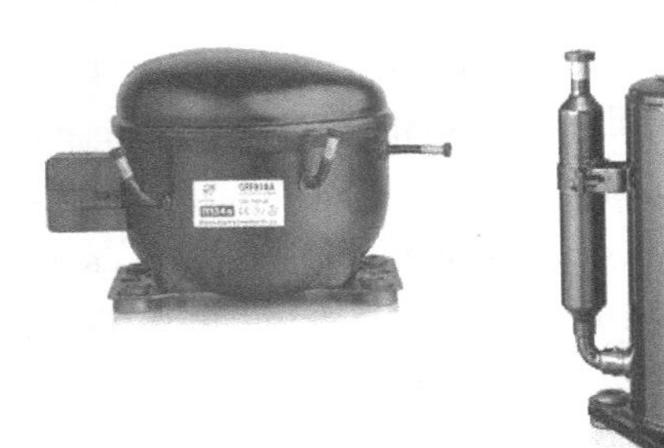

Fig. 1. 12-2 Hermetic compressors

Semi-hermetic compressors are constructed from castings that are bolted together using gaskets to form a hermetic seal. An electric drive motor is enclosed within the hermetically sealed crankcase so no external shafts or seals are required. Semi-hermetic compressor are used in medium and large capacity applications. Fig. 1. 12-3 is a typical semi-hermetic compressor.

Compressors may also be divided into positive displacement and dynamic types. Positive displacement types include reciprocating and rotary compressors. Positive displacement compressors compress volumes of low-pressure gas by physically reducing the volumes to cause a pressure increase, whereas dynamic types raise the velocity of the low-pressure gas and subsequently reduce it in a way that causes a pressure increase. Fig. 1. 12-4 shows the classification of the refrigeration compressors.

The reciprocating piston type is widely used. Automatic pressure-actuated suction and discharge valves are used. As the piston moves in the suction stroke, the suction valve opens to admit gas from the evaporator. When the suction stoke ends and the piston is at the bottom of the cylinder, suction valve will close and then the compression stroke begins. When the cylinder

Fig. 1. 12-3 A typical semi-hermetic compressor

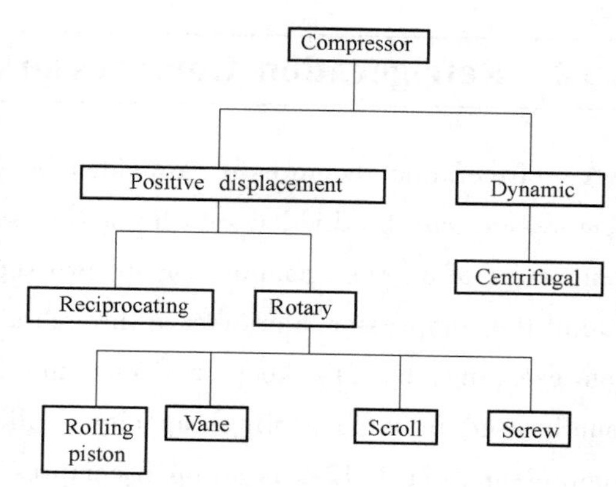

Fig. 1. 12-4 Classification of refrigeration compressors

pressure is higher than that in the discharge pipe, the discharge valve opens and the compressed gas enters the condenser.

The swash plate compressor, which is widely used in automobiles, is one kind of reciprocating compressor. The structure is illustrated in Fig. 1. 12-5. The pistons move back and forth when the shaft and swash plate rotate by power. The ports of refrigerant intake and discharge locate at the ends of the piston.

The rotary compressor includes screw, scroll, vane and rolling piston types. The screw compressor can be seen as a development of the gear pump. For gas pumping, the rotor profiles are designed to give maximum swept volume and no clearance volume. The screw compressors can be divided into twin screw and single screw types. Fig. 1. 12-6 shows a twin screw compressor. It can be seen that there are two mating rotors. The male rotor is typically turned by the compressor's drive motor. The female rotor is driven by the turning motion of the male rotor. The pitch of the helix is such that the inlet and the outlet ports can be arranged at the ends instead of at the sides. The solid portions of the screws slide over the gas ports to separate one stroke from the next so that no inlet or outlet valves are needed.

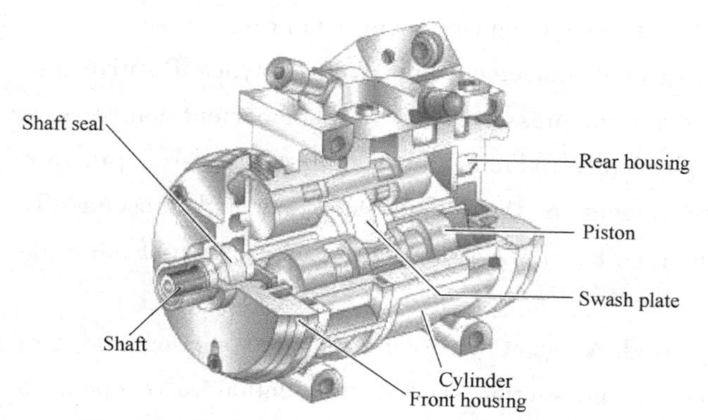

Fig. 1. 12-5 A Toyota swash plate compressor

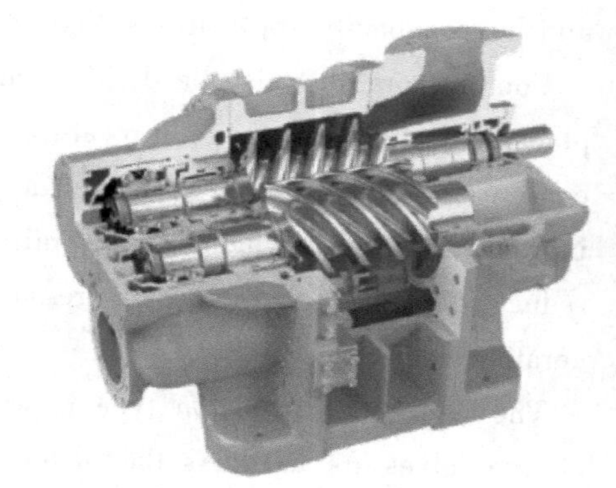

Fig. 1. 12-6 A typical twin screw compressor

Scroll compressor as shown in Fig. 1. 12-7 is positive displacement machine that compresses refrigerants with two spiral-shape scrolls.

The top scroll remains stationary while the bottom one is driven by the drive motor moving in an orbit around a fixed point in the stationary scroll. The suction vapor enters the scroll unit when an opening forms along the perimeter edges of the scrolls. As the orbiting scroll turns, the volume between the orbiting and stationary scrolls becomes progressively smaller compressing the vapor. As the volume is being reduced, the vapor is pushed toward the center of the scrolls where the discharge port is located.

The rotary vane compressors are positive displacement compressors as shown in Fig. 1. 12-8. Note that the axis of rotor is not at the center of cylinder. There are several radial slots in rotor. A radial sliding vane is installed in each radial slot. As the rotor rotates eccentrically, the vanes are thrown out from the slots by centrifugal force and the tips of the vanes contact the inner wall of the cylinder tightly. The volumes between an eccentric rotor and sliding vanes vary with angular position.

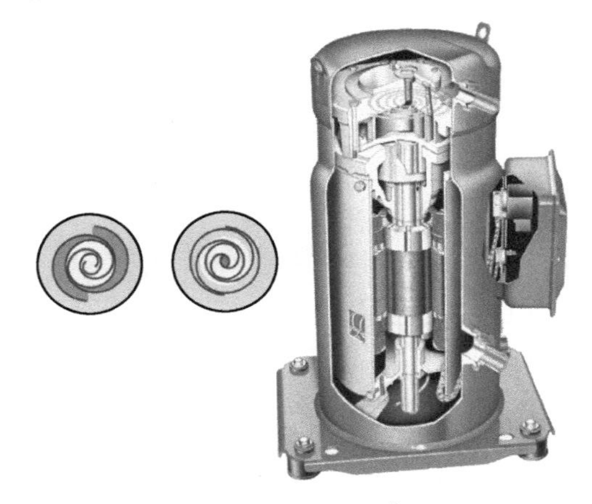

Fig. 1. 12-7 A typical scroll compressor

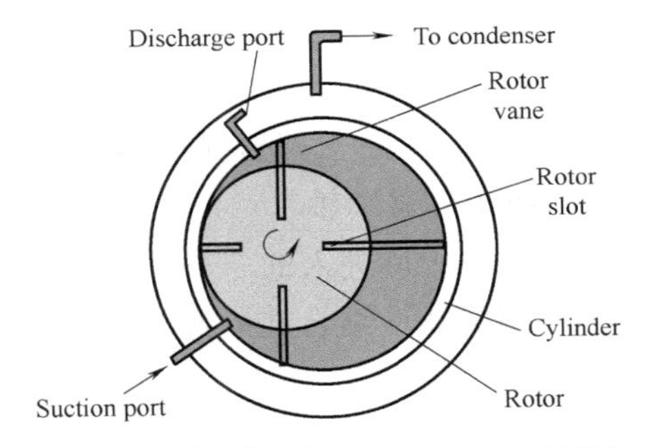

Fig. 1. 12-8 A rotary vane compressor

Larger models have eight or more blades and do not require inlet or outlet valves. The sealing is improved by the injection of lubricating oil along the length of the vanes. Rotary vane machines have no clearance volume, but they are limited in application by the stresses set up by the thrust on the tips of the vanes.

Rolling-piston compressors have one or two blades, which do not rotate, but are held by springs against an eccentric rotating roller. These compressors require discharge valves. This type has been developed extensively for domestic appliances, packaged air-conditioners and similar applications, up to a cooling capacity of 15kW. The working principle of a rolling-piston compressor is shown in Fig. 1. 12-9.

Dynamic compressors transfer energy to the gas by velocity or centrifugal force and then convert it to pressure energy. The most common type is the centrifugal compressor that suction gas enters axially into the eye of an impeller, which has curved blades, and is thrown out tangentially from the blade circumference. Centrifugal compressor is dynamic type. Fig. 1. 12-10 is a water chiller in which a centrifugal compressor is mounted on the upper part.

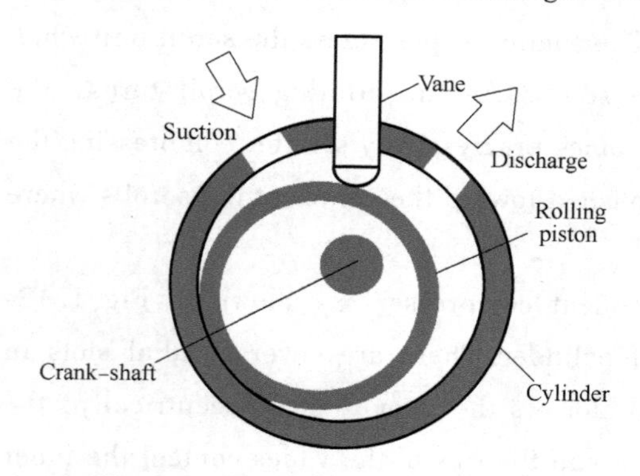

Fig. 1. 12-9 Working principle of a rolling-piston compressor

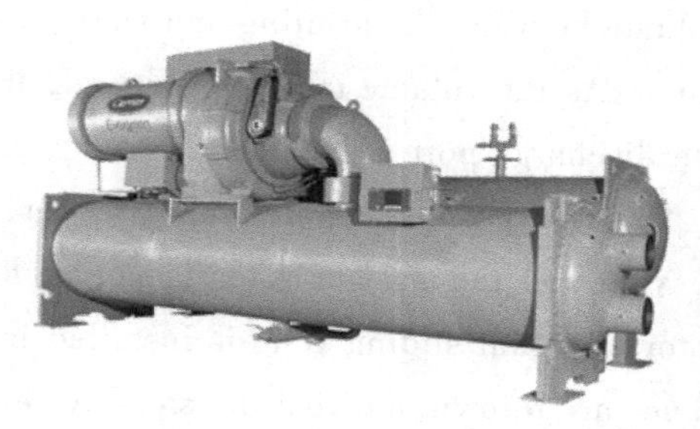

Fig. 1. 12-10 A centrifugal chiller

词汇表

intake 吸气
hermetic compressor 全封闭压缩机
semi-hermetic compressor 半封闭压缩机
open compressor 开启式压缩机
driveshaft 驱动轴
accessible 可进入的
coupled with 结合在一起
medium 中等的
enclose 装入，放入封套
positive displacement compressor 正排量容积式压缩机
dynamic 动力的
dynamic compressor 速度式压缩机
rotary compressor 回转式压缩机
subsequently 随后
piston 活塞
stroke 行程
automatic pressure-actuated suction and discharge 压力传感自动控制吸排气
cylinder 气缸
swash plate compressor 斜盘式压缩机
screw compressor 螺杆式压缩机
scroll compressor 涡旋式压缩机

vane 叶片
rolling piston 滚动活塞
gear pump 齿轮泵
profile 外形
swept volume 扫过的容积
twin screw compressor 双螺杆压缩机
single screw compressor 单螺杆压缩机
mating rotors 相互啮合的转子
male rotor 阳转子
female rotor 阴转子
helix 螺旋，螺旋状物
pitch 螺距
pitch of a helix 螺杆螺距
spiral-shape 螺旋形
stationary 静止的
orbit *n*. 轨道；*v*. 绕轨道运行
fixed 固定的
perimeter 周长
edge 边缘
eccentric rotor 偏心转子
sliding vane 滑片
angular 角度的
centrifugal 离心的
thrust 推力
rotary vane compressor 旋转叶片式压缩机
rolling-piston compressor 滚动活塞式压缩机
spring 弹簧
impeller 叶轮
tangentially 沿切线地
mount 安装

学习要点

连词主要分为两类：并列连词和从属连词。并列连词用来连接平行的词、词组或分句。从属连词是用来引起从句的。

1. 并列连词

常见的并列连词有 and，but，or，nor，so，yet，for，hence，as well as，both … and，not only … but also，either … or，neither … nor，then 等。

并列连词中有些是表示意思转折的，如 but。

例 1.12-1 Rotary vane machines have no clearance volume, but they are limited in application by the stresses set up by the thrust on the tips of the vanes.

2. 从属连词

常见的从属连词有 after, when, before, as, while, since, until, till, although, though, if, even if, unless, because, than, whether, so that, as soon as, as long as, in order that, as if, suppose (that), provided (that), in case (that) 等。

例 1.12-2 As (when) the volume is being reduced, the vapor is pushed toward the center of the scrolls where the discharge port is located.

该例句中 as 或 when 引导的从句是时间状语从句。

例 1.12-3 The volumetric efficiency of the compressor is further improved because there is no clearance volume required in this type of compression.

该例句中的 because 引出时间状语从句。

练习

1. 请阅读下面一段英文，并翻译为汉语。

There are no suction or discharge valves required in a scroll compressor so the flow of vapor through the compressor is continuous. The smooth flow of vapor through the compressor eliminates much of the vapor pulsations, flow losses and noise associated with valved systems. The separation of suction and discharge locations substantially reduces heat transfer between these vapors, improving the volumetric efficiency. The volumetric efficiency of the compressor is further improved because there is no clearance volume required in this type of compression. The energy efficiency of the scroll compressor is relatively high, with full load efficiencies nearly equal for reciprocating and scroll compressors. Part-load scroll efficiencies are 10% to 20% higher than those of reciprocating compressors. Since the unit operates at part load most of its useful life, the energy savings are very high for scroll compressors [5].

2. 请翻译下列句子。

1) Reciprocating compressors convert rotational motion produced by an electric motor into reciprocating motion that alternately increases and decreases the working volume of its cylinders.

2） A rolling-piston compressor incorporates a cylindrical steel roller that is turned by an eccentric shaft.

3. 请翻译下列句子。

1） 二氧化碳活塞压缩机主要应用于热泵和食品冷冻冷藏运输领域。

2） 全封闭活塞式压缩机主要用于家用冰箱和冷柜。

3） 涡旋式压缩机可以用于家用空调器。

4. 请用 as/when/if 填空。

1） ________ the piston moves beyond its top-dead-center, the suction stroke begins.

2） ________ the piston reaches the lowest point in its cycle, the volume of the cylinder reaches its maximum value.

3） ________ water flows through a pipe with a velocity V, the following characteristics are observed by injecting dye as shown.

1.13 Automatic Control in Refrigeration and Air Conditioning Systems

Automatic control is necessary and important in refrigeration and air conditioning systems. They have three main functions:

1) Keep the temperature, humidity, pressure and air quality in the conditioned space at desired value.

2) Make the system operate safely.

3) Save energy by improving control strategy.

A typical control process can be described with a block diagram as shown in Fig. 1.13-1. A typical control process includes a sensor, a controller, a controlled device and a controlled plant. The desired value is set at comparison element in controller. The controlled variable is measured by the sensor. The difference between measurement and set value is called error.

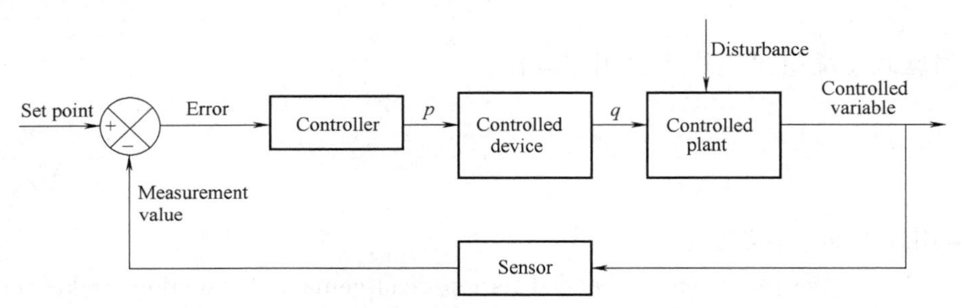

Fig. 1.13-1 A typical control process

The error signal enters controller. The controlled device includes an actuator and a connected valve or damper. The output signal of the controller enters an actuator. If the error is not zero, the controller generates an output signal based on a certain control logic that is sent to the actuator to make it move. The opening of connected valve or damper increase or decrease with movement of the actuator, which causes flow rate of working fluid to be changed to regulate controlled variable. Disturbance and the regulating effect act on the controlled plant together to effect the controlled variable. As a result, the controlled variable is adjusted and the error decreases. This control loop repeats and repeats until the error is zero or near zero and the controlled variable is equal to the desired value or near desired value.

The output of this control system is the controlled variable that is measured by the sensor. The output signal of the sensor is sent to input end of the control system to be compared with the set value. This process is called feedback. Normally the automatic control used in refrigeration and air conditioning system is a closed loop and negative feedback control system.

Different controller has different control logic. According to the relation of output and input signal, the controllers can be divided into the following types: on-off, proportional (P), integral (I), proportional-integral (PI) and proportional-integral-derivative (PID) etc. Fig. 1. 13-2 shows the input and output signal of a controller. The input signal of controller is $e(t)$ and the output signal is $m(t)$.

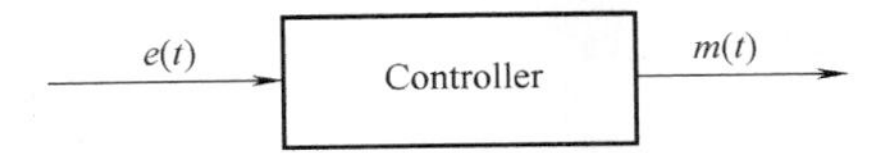

Fig. 1. 13-2 Input and output signal of a controller

The control logic can be written in the following equations for different types of controllers respectively:

1) On-off controller: $\begin{cases} M_1, e(t)>0 \\ M_2, e(t)<0 \end{cases}$

2) Proportional controller (P): $m(t)=K_p e(t)$

3) Integral controller (I): $m(t)=K_i \int_0^t e(t)\,dt$

4) Proportional-integral controller (PI): $m(t)=K_p e(t)+K_i \int_0^t e(t)\,dt$

5) Proportional-integral-derivative controller (PID):

$$m(t)=K_p e(t)+K_i \int_0^t e(t)\,dt+K_d \frac{de(t)}{dt}$$

For a proportional controller (P), the output signal of controller $m(t)$ is proportional to the input signal of controller $e(t)$. For a proportional-integral controller (PI), the output signal $m(t)$ is equal to $K_p e(t)+K_i \int_0^t e(t)\,dt$. It can be seen that the name of the controller reflects the relation of output and input signal. The controllers have different appearances as shown in Fig. 1. 13-3.

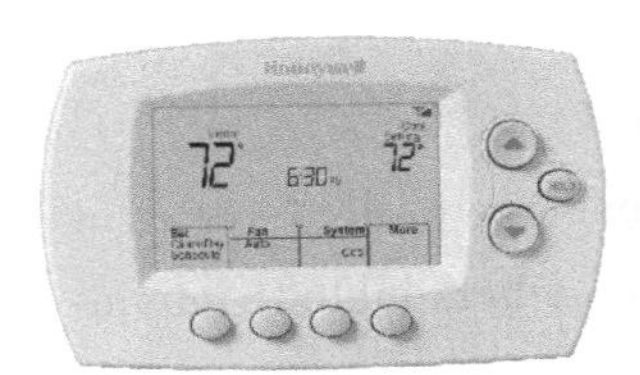

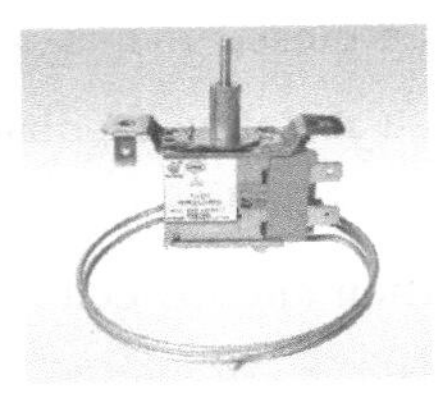

Fig. 1. 13-3 Various controllers

The controlled device consists of an actuator and a connected valve or damper as shown in Fig. 1. 13-4. The actuator, which can be electric, pneumatic or mechanical, is connected with a valve or a damper. The movement of actuator is controlled by output signal of the controller.

The sensors are also very important for a control system. The respond speed of the sensor must be faster than the rate of change of the controlled variable. Resistance temperature detector (RTD), thermistor and thermocouples are common temperature sensors as shown in Fig.1.13-5. Typical pressure sensors are shown in Fig. 1. 13-6.

An automatic control system can be so summarized:

a)

b)

Fig. 1. 13-4 Controlled devices

a) Actuators and valves b) Actuators and dampers

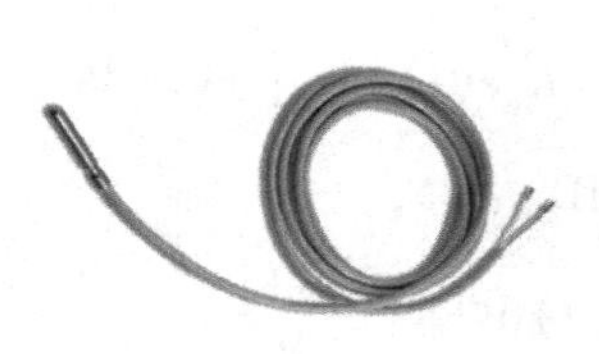

a)

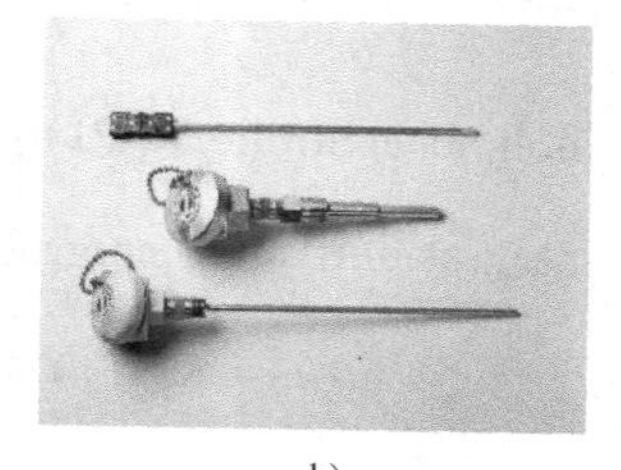

b)

Fig. 1. 13-5 Various temperature sensors

a) RTD b) Thermocouples

Fig. 1. 13-6 Typical pressure sensors

1) No person directly participates in a control process.

2) The control loop consists of sensor, controller, controlled device and controlled plant.

3) It is a closed loop system.

4) It is a negative feedback control system.

Temperature and pressure are the most common controlled variables in refrigeration and air conditioning systems. The others include humidity, liquid level etc.

A typical example using control in refrigeration system can be found in a refrigerator. On-off control is frequently-used in refrigerators because it is cheap, simple and reliable. It provides only two signals: on and off. Theoretically, the output signal is "on" if the error is greater than zero and "off" if the error is smaller than zero. On-off controllers will be switched frequently and damaged quickly due to the random disturbance. To handle this problem, a differential range is designed for each on-off controller as shown in Fig. 1. 13-7. In other words, there are an upper limit and a lower limit for the controlled variable for each on-off control. On or off switches only when the error is so large that exceeds the limit of the differential range. If the error is smaller than the limit of the differential range, the state of "on" or "off" will remain unchanged. The life time of an on-off controller gets improved in this way.

The temperature in the refrigerator is usually controlled by an on-off controller. The temperature-regulating curve for this on-off control used in refrigerator can be seen in Fig. 1. 13-8.

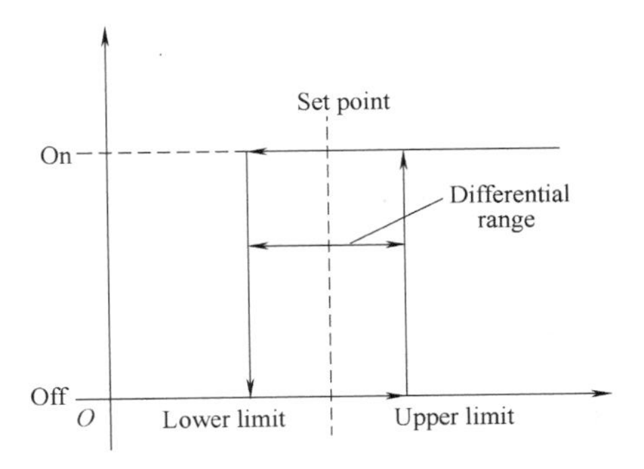

Fig. 1. 13-7 Differential range

When the temperature is higher than the upper limit, the controller closes the electric circuit of the compressor automatically. The refrigeration system works and the temperature declines continuously. As the temperature reaches the lower limit, the controller breaks the circuit of the compressor. The refrigeration system stops working and the temperature increases gradually. As the temperature is over the upper limit, the compressor runs again. This regulation process repeats and maintains the controlled variable around the set point.

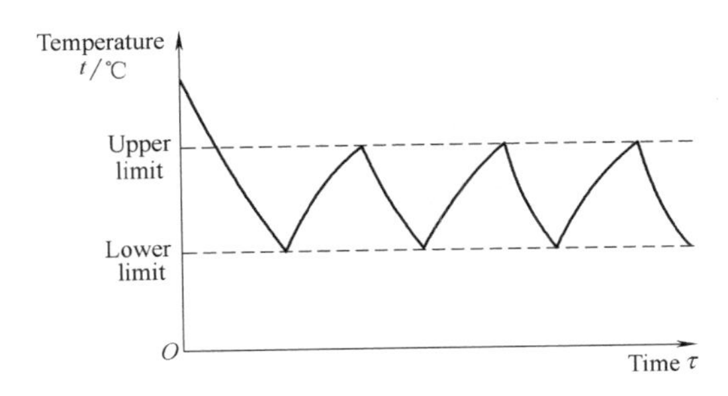

Fig. 1. 13-8 Temperature-regulating curve for an on-off control refrigerator

词汇表

automatic control 自动控制
operate 运行
control strategy 控制策略
control loop 控制回路
sensor 传感器，发信器
controller 控制器，调节器
controlled device 执行器
controlled plant 调节对象
desired value 理想值
comparison component 比较元件
set 设定
controlled variable 被调参数
measure 测量

error 偏差
signal 信号
actuator 执行机构
damper 风门
control logic 控制逻辑
opening 开度
flow rate 流量
disturbance 干扰
output 输出
input 输入
feedback 反馈
closed loop 闭环
negative feedback 负反馈
on-off controller 双位调节器，通断控制器
proportional controller 比例调节器
integral controller 积分调节器
proportional-integral controller 比例积分调节器
proportional-integral-derivative 比例积分微分调节器
resistance temperature detector（RTD） 热电阻
thermistor 热敏电阻
thermocouples 热电偶

学习要点

1. 情态动词

情态动词主要有 can（could）、may（might）、must、ought、need、dare（dared）。此外，shall、will、should、would 在一定场合下也可用作情态动词。情态动词加动词原形一起构成谓语。情态动词本身带有一定的词义。

can 表示什么事是可能的或是允许的，或某人有能力做某事。

could 是 can 的过去式。

must 表示必须、一定。

may 或 might 表示可能。

should 表示应该。

would 表示当我们想象发生某事时，将……

2. 情态动词+动词的被动语态

科技论文中经常使用情态动词加动词的被动语态形式。

例 1.13 A typical control process can be described with a block diagram as shown in Fig. 1. 13-1.

一个典型的控制系统可以用图 1.13-1 所示的框图描述。

练习

1. 英汉互译。

1）室温可以通过设定空调器理想温度调节。

2）冰箱通常使用双位调节器控制冰箱内的温度。

3）A thermostat is a combination of a temperature sensor and a controller.

4）All elements of the feedback loop can effect control performance.

5）The purpose of any closed-loop controller is to maintain the controlled variable at the desired setpoint.

2. 请用情态动词 can，may，must 填空。

1）__________ I come in ?

2）__________ you follow her talk?

3）Hotel __________ be booked at least one week before departure.

4）Compressors __________ be classified into two types.

1.14 Distribution Systems

The air temperature, humidity and cleanliness can be maintained at desired value by distributing the conditioned air (heated, cooled, humidified, dehumidified) or chilled water to every conditioned room through ducts or pipes. Systems of ducts, known as ductwork, are a central component of a building´s heating, ventilation and air conditioning (HVAC) system. In most systems, only one set of ductwork is present, which is used to supply fresh air to satisfy the requirements of human health. The split air conditioners are ductless.

A fan coil unit (FCU) is a simple device that consists of heating or cooling coil and a fan, which is one kind of HVAC system. Fan coil units can be found in residential, commercial and industrial buildings. The cooling or heating load in room can be handled by chilled or hot water that flows through coils. Sometimes fan coil units combine with outdoor air ductwork for fresh air requirements.

A typical air distributing system consists of air ducts, fittings, valves and fans. Similarly, a typical water pipe distributing system includes water pipes, fittings, valves and pumps. One important task of an air conditioning system design is hydraulic calculation to determine the size of air ducts or water pipes and select the suitable fans or pumps.

Calculating size of air ducts belongs to ventilation system design. There are many different methods. The most common used ways are:

1) Velocity reduction method (suitable for residential or small commercial installations).

2) Equal friction method (suitable for medium to large sized commercial installations).

3) Static regain (suitable for very large installations such as concert halls, airports etc.)

Equal friction method is the most common method used for commercial HVAC systems.

The procedures of ductwork design can be summarized as following:

1) Calculate the hourly load for each room and find the peak load of each room.

2) Calculate the air mass flow rate for each room.

3) Sketch out the ductwork route onto the floor plan.

4) Determine the size of ducts through velocity reduction, equal friction or static regain methods.

5) Calculate fitting losses.

6) Add up all the pressure losses from the start to the exit of each branch.

7) Find the run which is with the largest pressure drop. It's usually the longest run but could also be the run with the most fittings.

8) Select the suitable fan which must overcome the largest loss.

9) Add dampers to the branches to ensure equal balance pressure drop to achieve the design flow rates to each room.

When fan coil units are used for air conditioning, following procedures are suggested.

1) Calculate the peak load for each room.

2) Select suitable fan coil units and determine the water flowrate for fan coil unit.

3) Draw all pipe routing, piping accessories and fan coil units.

4) Determine the water velocity in piping or average specific friction resistance.

5) Determine the size of piping using related charts or software.

6) Sum all the pressure drops for each loop.

7) Add balance valves to solve the hydraulic imbalance.

8) Find the loop which has the largest pressure drop.

9) Select the suitable pump which must overcome the largest pressure drop.

More commonly used air conditioning system is a system that combines the above-mentioned two systems. That is called independent fresh air system plus fan coil units. Both duct and pipe sizing must be done for this kind of system.

词汇表

duct 管道，风道

pipe 管道，圆形管道

distribute 分布，分配

humidified 加湿

dehumidified 去湿

ductwork 风道系统

split air conditioners 分体空调器

ductless 无风道的

fan coil unit 风机盘管机组

handle 处理

fittings 管道配件，管接头

hydraulic calculation 水力计算

velocity reduction method 假定速度法

equal friction method 压损平均法

static regain 静压复得法

hourly load 逐时负荷

peak load 峰值负荷

air mass flow rate 空气质量流量

sketch out 画出（草图）

ductwork route 管道线，管线

floor plan 平面图

add up 加起来

branch　支路
overcome　克服
damper　风阀
pressure drop　压降
water flowrate　水流量
pipe routing　管道布置
piping accessories　管件
average specific friction resistance　平均比摩阻
chart　图表
sum　求和
loop　回路
balance valve　平衡阀
solve　解决
hydraulic imbalance　水力不平衡
pipe sizing　确定管道尺寸

学习要点

1. 过去分词作定语

例 1.14-1　The air temperature, humidity and cleanliness can be maintained at desired value by distributing the conditioned air (heated, cooled, humidified, dehumidified) or chilled water to every conditioned room through ducts or pipes.

该例句中 desired 是过去分词作定语修饰 value, conditioned air 是过去分词 conditioned 作定语修饰 air, chilled water 是过去分词 chilled 作定语修饰 water, conditioned room 中是过去分词 conditioned 作定语修饰 room。

2. 动词+ing

动词+ing 有时是动名词，有时是现在分词。动名词主要起名词作用，现在分词主要起形容词和副词的作用。

例 1.14-2　Systems of ducts, known as ductwork, are a central component of a building's heating, ventilation and air conditioning (HVAC) system.

该例句中的 building、heating、air conditioning 都是动名词。

例 1.14-3　When fan coil units are used for air conditioning, following procedures are suggested.

该例句中的 following 是动词 follow 的现在分词形式，在句中的作用相当于形容词。

例 1.14-4　Determine size of piping using related charts or software.

该例句中 using 是动词 use 的现在分词形式，using related charts or software 是现在分词短语作状语，说明用什么样的方式确定管径大小。

练习

1. 请将下列短文翻译为汉语。

Types of Piping Systems

Before piping design can be discussed in detail, you must first have an understanding of the three basic type of piping systems: closed loop, open loop and once-thru.

Closed-Loop (Evaporator)

In a closed-loop piping system, the water is contained within a closed piping system, or loop, through which it circulates. Typically, closed loop systems are chemically treated to control corrosion, scale, slime and algae within the piping but their chemical treatment requirements are not as extensive as an open loop.

Open-Loop (Condenser)

In an open-loop system, the water is in constant contact with the air and the system is therefore open to the atmosphere. A typical example of an open-loop system is a recirculating condenser water system with a cooling tower, sprayed over the tower media surface, collected into the tower basin, circulated through the condenser, and then sent back through the cooling tower.

Once-Thru

In this type of system, water passes through the system once and is then discharged. An example of a once-thru system would be a chiller with river water piped into its water-cooled condenser. The rejected heat from the condenser is introduced back into the river, which is not always acceptable from an environmental perspective. In general, once-thru systems that use "city water" are not allowed because they use excessive amounts of water [6].

2. 请用英语简单介绍一下上题中提到的三种管道系统。

3. 请重写下面的句子，用定语从句代替下句中的过去分词。

A typical example of an open-loop system is a recirculating condenser water system with a cooling tower, sprayed over the tower media surface, collected into the tower basin, circulated through the condenser, and then sent back through the cooling tower.

1.15 Flow in Pipes

The air conditioning systems contain all-air, air-water and all-water systems. Air flows in ducts in an all-air system. Water flows in water pipes in an all-water system. There are both air ducts and water pipes in an air-water system. In a vapor compression refrigeration system, refrigerant flows in sealed pipes.

Fluid includes gas and liquid. Air, water and refrigerant are all fluid. A typical water pipe system includes pumps, pipes, fittings and valves. The fluid flow in a pipe may be laminar, turbulent or transitional flow. Osborne Reynolds (1842-1912), a British scientist and mathematician, was the first to distinguish the difference between these classifications of flow by using a simple apparatus as shown in Fig. 1.15-1. If water flows through a pipe with a velocity V, the following characteristics are observed by injecting dye as shown. If the water flowrate is small enough, the dye streak will remain as a line. This kind of flow is called laminar flow. If the flowrate is large enough, the dye streak is not a line but a curve and almost immediately becomes blurred. This kind of flow is called turbulent flow. If the flowrate is intermediate, the dye streak fluctuates in time and space and appears irregular behavior intermittently. This kind of flow is called transitional flow.

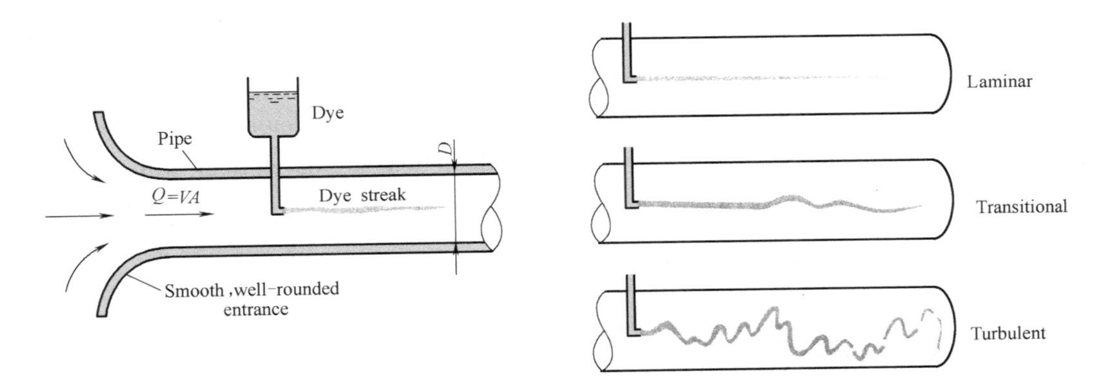

Fig. 1.15-1 Experiment to illustrate type of flow

The distinction of different flow types and its dependence on dimensionless number Re was first proposed by Osborne Reynolds in 1883. Re is defined as the ratio of inertia to viscous effects in the flow as shown in Eq. (1.15-1).

$$Re = \frac{\rho VD}{\mu} \tag{1.15-1}$$

where ρ is density, V is average velocity in the pipe, D is diameter of the pipe and μ is dynamic viscosity of fluid.

The Reynolds number ranges for laminar, transitional or turbulent pipe flows are difficult to be given precisely. Normally the flow in a round pipe is laminar if *Re* number is less than 2100. The flow in a round pipe is turbulent if *Re* is greater than 4000. For *Re* numbers between these two limits, the flow is transitional flow.

A typical pipe system usually consists of many straight pipes, valves and elbows etc. The overall head loss for the pipe system includes the major loss that is the head loss due to viscous effects in the straight pipes and the minor loss which is caused by valves, tees, elbows etc.

The major loss can be calculated with the following Darcy-Weisbach equation.

$$h_{L\ major} = f\frac{l}{D}\frac{V^2}{2g} \tag{1.15-2}$$

where f is friction factor, V is average velocity, l is pipe length and D is diameter of the pipe.

For laminar fully developed flow, the value of f is equal to $64/Re$. For turbulent flow, f is equal to $\phi\left(Re, \frac{\varepsilon}{D}\right)$. ε is roughness of the pipe wall. $\frac{\varepsilon}{D}$ is relative roughness. Eq. (1.15-2) is valid for any fully developed, steady, incompressible pipe flow.

Moody chart shows the functional dependence of f on Re and $\frac{\varepsilon}{D}$. The following equation from Colebrook is valid for the entire non-laminar range of the Moody chart.

$$\frac{1}{\sqrt{f}} = -2.0\lg\left(\frac{\frac{\varepsilon}{D}}{3.7} + \frac{2.51}{Re\sqrt{f}}\right) \tag{1.15-3}$$

Eq. (1.15-3) is an empirical formula, which is called Colebrook formula. Note that this equation is implicit. Therefore, it is not easy to be used by hand calculation because it is not possible to solve f without iteration. However, such calculations are not difficult with computers.

The pressure drop from cross section 1 to cross section 2 can be written as

$$\Delta p = p_1 - p_2 = \gamma(z_2 - z_1) + \gamma h_L = \gamma(z_2 - z_1) + f\frac{l}{D}\frac{\rho V^2}{2} \tag{1.15-4}$$

where $\gamma = \rho g$ is the specific weight of the fluid, z_1 is vertical elevation of axis at cross section 1 and z_2 is vertical elevation of axis at cross section 2.

The minor losses which are caused by valves, bends and tees can be calculated with the following equation.

$$h_{L\ minor} = K_L\frac{V^2}{2g} \tag{1.15-5}$$

where K_L is loss coefficient, which is defined as

$$K_L = \phi(\text{geometry}, Re) \tag{1.15-6}$$

The actual value of K_L is strongly dependent on the geometry of the component considered. For many practical applications, the *Re* number is large enough so that the flow through the component is dominated by inertia effects and the importance of viscous effects is secondary. In most cases, the loss coefficients for components are function of geometry only. The loss coefficients for pipe components can be found in HVAC manuals.

The pressure drop across a component can be written as

$$\Delta p = K_L \frac{1}{2}\rho V^2 \tag{1.15-7}$$

In designing an air-water air conditioning system, the water pipe systems and air ducts system must be calculated with these above-mentioned equations.

词汇表

all-air system 全空气系统
all-water system 全水系统
air-water system 空气-水系统
sealed 密封的，封闭的
fitting 配件
laminar flow 层流
turbulent flow 湍流
transitional flow 过渡流
Osborne Reynolds （人名）英国科学家雷诺
observe 观察
inject 喷射
dye 染料
streak 条纹，线条
blur 变模糊
intermediate 中等的
intermittently 间歇地
dimensionless number 无量纲数
ratio 比率，比
inertia 惯性
viscous 黏性的
dynamic viscosity 动力黏度
Reynolds number 雷诺数
elbow 弯头
head loss 压头损失
major loss 沿程损失
minor loss 局部损失
tee 三通
Darcy-Weisbach equation 达西-魏斯巴赫公式
friction factor 摩擦系数

fully developed flow　充分发展流
roughness　粗糙度
pipe wall　管壁
steady　稳态的
incompressible　不可压缩的
Moody chart　莫迪图
functional　函数的
non-laminar　非层流的
empirical formula　经验公式
implicit　隐式的
iteration　迭代
cross section　截面
specific weight　比重
vertical elevation　垂直高度
axis　轴心
bend　弯管
loss coefficient for component　部件或管件阻力系数
dominate　占优势，占主要地位
secondary　次要的
geometry　几何尺寸
manual　手册
component　部件，管件

学习要点

连接副词

连接副词主要用来连接句子和句子，或者连接子句和子句，以使句子意思连贯。常见的连接副词有 therefore，otherwise，however，moreover，consequently，finally 等。

例 1.15-1　Note that this equation is implicit. Therefore, it is not easy to use by hand calculation because it is not possible to solve *f* without iteration. However, such calculations are not difficult with computers.

该例句中 therefore 表示所以，however 是然而的意思，表示意思转折。

例 1.15-2　This throttling process causes some of refrigerant to flash. Consequently, the temperature of the remaining liquid decreases.

节流过程引起一些制冷剂闪发，结果余下的制冷剂温度下降。

该例句中 Consequently 是连接副词，连接两个句子，使句子意思连贯，并表达出上下句间的因果关系。

练习

1. 英汉互译。

1）达西-魏斯巴赫公式可写为：$h_{L\ major}=f\dfrac{l}{D}\dfrac{v^2}{2g}$，其中$h_{L\ major}$是沿程损失，$f$是摩擦系数，$l$是管长，$D$是管径，$v$是流体的平均流速，$g$是重力加速度。

2）管内流动可分为层流、湍流和过渡流。

3）雷诺数是惯性力和黏性力的比值。

4）The loss coefficients for valves, tees and elbows can be found in HVAC manuals.

2. 请使用下列连接副词各写一个句子。

therefore otherwise however moreover consequently finally

1.16 Cooling Towers

The condensers in vapor compression refrigeration systems can be air cooled or water cooled. If a condenser is water cooled, a cooling tower must be installed for removing heat from high temperature refrigerant vapor in condenser by the cooled water flowing through the condenser. By spraying hot water from condenser, some water vaporizes and the remaining water is cooled in tower. The cooled water is sent back to condenser by pump to absorb heat of condensation. Cooling towers can also be seen in absorption systems for removing heat from condensers and absorbers.

A typical HVAC/R cooling tower consists of a galvanized, plastic or wood enclosure that contains the hot water distribution system, fill, fan system, cold water basin, make-up water system, piping connections, access ladder, decks and control systems. Cooling towers are normally installed on a concrete pad for ground units or suspended on a steel frame for roof-mounted units.

The air flow in tower can be natural, forced or induced draft. A natural-draft tower employs the difference in density between the entering and leaving air to establish natural convection through the tower. Normally this kind of tower must be very high and large to be effective. Forced-draft towers have centrifugal fans that propel the air entering and flowing through the tower. Induced-draft towers draw the saturated air out of the tower using a large axial fan. In all these designs, the hot water enters the tower at the top and flows by gravity to the bottom where the cooled water is collected.

For even distribution of hot water, nozzles are usually used to spray water over the fill. In this type of tower, the hot water from condenser is pumped to the top of the tower where it is sprayed through a series of nozzles. To cool the hot water effectively, the evaporation of water in tower need to be enhanced. Therefore, the water surface exposed to the atmosphere must be maximized. Since the amount of exposed water surface depends primarily on the spray pattern, a good spray pattern is essential to high efficiency. A hot water basin perforated with a pattern of orifices in its bottom sheet is another type of distribution design. But the nozzles and orifices can be blocked by scale and debris. Therefore, strainers and water treatment are necessary to reduce block.

When the hot water is distributed across the tower, it encounters the air current and breaks up into droplets. As the outer surface of the droplets evaporates, the water in the droplet center is cooled. To further increase evaporation rate, fill is added to the tower. Fill is any material that is used to increase the exposed surface area of water and the time length that water is exposed to the air. Fill is formed into a shape that encourages the water to flow in a thin film and has an extremely large surface area. As the water spreads over the surface of the fill, the evaporation rate increases.

Tower fill can be made with PVC plastics. The location of the fill and its orientation depends on the convection in the tower. Natural-convection towers typically use slat-type PVC fill. Honeycomb PVC fill is used in induced-draft tower. Forced-draft cooling towers use high static pressure fans to blow air from base or side of the enclosure through PVC slats.

In forced-draft tower designs, centrifugal blowers are used to develop the high velocities and static pressure needed. Induced-draft towers use large-diameter axial fans to move large quantities of air at lower velocity. The tower fan can be rotated by direct drive, belt drive, or using a drive shaft and gearbox.

After the hot water that enters the tower has been cooled, it is collected in the water basin at the bottom of the closure. The make-up water valve is located in the basin to maintain the desired level. An overflow pipe is installed in the basin to prevent water in the basin from rising too high. The excess water is directed to a drain. Steam piping and electric resistance heaters can be installed in the cold water basin to prevent freezing during cold weather.

词汇表

galvanized 镀锌的
enclosure 外壳
distribution system 分布系统，布液系统
fill 填料
basin 水池
make-up water system 补水系统
access ladder 通道竖梯
deck 甲板，平台
concrete pad 混凝土基座
ground unit 落地式机组
suspend 悬挂
roof-mounted units 屋顶安装式机组
steel frame 钢架
natural-draft tower 自然通风冷却塔
forced-draft tower 机械通风冷却塔
induced-draft tower 诱导通风冷却塔
centrifugal fan 离心风机
propel 推动
draw out 抽出
axial fan 轴流风机
gravity 重力

even distribution　均匀分布
nozzle　喷嘴
perforated　打孔的
orifice　孔
sheet　薄板
block　堵塞
scale　垢，水垢
debris　杂物，碎片
strainer　过滤器
water treatment　水处理
encounter　遇到
air current　气流
break up　碎裂
droplet　液滴
thin film　薄膜
PVC plastics　聚氯乙烯塑料
location　位置
orientation　方向
slat-type　片状
honeycomb　蜂窝状
static pressure　静压
blow　吹
direct drive　直接驱动
belt drive　带传动
drive shaft　驱动轴
gearbox　齿轮箱
blower　风机
level　液位
overflow pipe　溢流管
drain　下水道
electric resistance　电阻

学习要点

1. 复杂长句分析

例 1.16-1　If a condenser is water cooled, a cooling tower must be installed for removing heat from high temperature refrigerant vapor in condenser by the cooled water flowing through the condenser.

该例句是一个复合句。主句为 a cooling tower must be installed for removing heat from high temperature refrigerant vapor in condenser by the cooled water flowing through the condenser。从句为时间状语从句 if a condenser is water cooled。

主句中主语为 a cooling tower，must be installed 为谓语，for removing heat from high temperature refrigerant vapor in condenser by the cooled water flowing through the condenser 为目的状语，high temperature refrigerant vapor 为介词 from 的宾语，in condenser 为动词 remove 的状语，by the cooled water flowing through the condenser 是动词 remove 的状语，flowing through the condenser 是现在分词短语作定语修饰前面的 the cooled water。

2. 现在分词作定语

例 1.16-2 例 1.16-1 例句中 flowing through the condenser 是现在分词短语作定语修饰前面的 the cooled water。此处也可以用定语从句表达，改写为：

If a condenser is water cooled, a cooling tower must be installed for removing heat from high temperature refrigerant vapor in condenser by the cooled water that flows through the condenser.

可以看出用分词表达更简洁。

练习

1. 请将下列句子翻译为英语。

1）冷却塔可分为自然通风冷却塔、强制通风冷却塔和诱导通风冷却塔。

2）冷却塔常用的风机为离心风机和轴流风机。

3）冷水塔底部的水池安装有补水管和溢水管。

4）吸收式制冷系统常使用冷却塔以带走吸收热和冷凝热。

2. 请试对下列句子进行句子成分分析，并翻译为汉语。

1） In forced-draft tower designs, centrifugal blowers are used to develop the high velocities and static pressure needed.

2） Because most of the heat transfer that occurs results from a phase change of liquid water into vapor, the primary atmospheric property governing the performance of a cooling water is the wet bulb temperature of the atmosphere entering the tower.

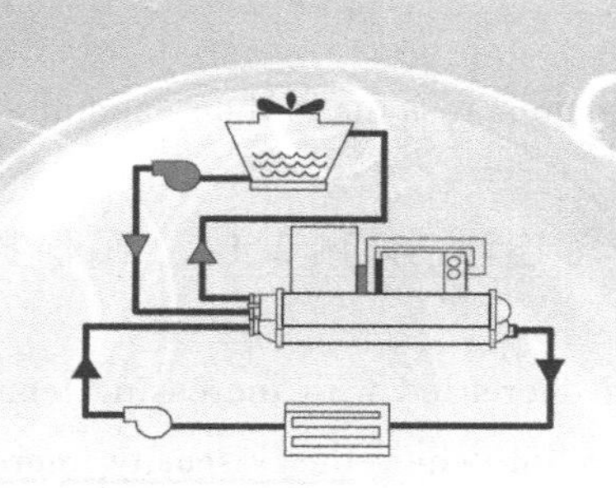

第2部分

制冷空调科技英语写作

2.1 英文科技论文与普通文章写作的区别

英文科技论文主要指发表在专业期刊上的原创研究论文、研究综述及各层次学位论文、研究报告等。制冷空调领域的研究人员、教师、学生、工程师等通常通过撰写学术论文、参加学术会议、阅读学术论文等方式进行学术交流，跟踪学术前沿，了解行业技术进展，启发创新思维。

英文科技论文的写作有许多不同于普通文章的特点，即与报刊上的新闻、小说、散文、书信等存在许多明显差异。这些不同表现在下列几个方面。

1. 读者不同，写法不同

科技论文的读者通常为同领域的学者、工程师、学生、技术人员等，同领域的读者对本领域的学科基本知识、术语是熟悉的，因此撰写制冷空调领域的学术论文时无须详细介绍诸如焓、熵等本学科的基本术语，无须像科普文章一样，用尽量浅显易懂的语言来解释和说明现象。写论文时要准确使用本领域的术语，用尽量简单、明确、客观的句子来论述自己的观点。这一点，对于中外论文的撰写都是一样的。

2. 内容不同

学术论文主要针对本领域学术研究的热点或空白，给出自己的研究方法和结论，体现出一定的创新性。

3. 写作风格不同

英文学术论文多使用较长的复合句，一句话浓缩很多内容。普通文章多使用简单句、并列句。

4. 文章的组织结构不同

学术文章的结构清晰。原创科技论文的写作，一般有比较固定的论文结构。

5. 语法上的特点

1）学术论文一般不用缩写形式，例如，does not 一般不写成 doesn't.

例 2. 1-1 This model does not consider the thermal resistance across the carbon nanotube-fluid interface.

2）使用的连词不同，学术论文常用 however，moreover 等连词，如例 2. 1-2。而非学术论文会用 only、anyway 等较口语化的词，如例 2. 1-3。

例 2. 1-2 Results showed that the viscosity of nanofluids decreased with increasing temperature obviously. However, the capillary tube diameter may influence the viscosity more strongly for higher nanoparticle mass fractions, especially at lower temperatures.

例 2. 1-3 I can't go to the theater, only I have to do experiments this evening.

3）学术论文经常使用动词的名词化形式。

例 2. 1-4 Commercial refrigeration focuses on the design, installation and maintenance of refrigeration equipment.

这里的 installation 和 maintenance 分别是动词 install 和 maintain 的名词化形式。

4）学术论文经常使用被动语态。

例 2. 1-5 Another interesting numerical investigation was conducted by Wang.

Thermal conductivity was measured using a KD 2 Pro thermal properties analyzer.

The model was found to be suitable for describing the fluid behavior.

注意这里没有用以 we 开头的主动语态。

5）学术论文经常使用简洁而正式的词汇，如例 2. 1-6。而非学术论文可采用较口语化的词汇，见例 2. 1-7。

例 2. 1-6 The nanofluids were assumed to be in single phase, in thermal equilibrium and without velocity slip between base fluid and particle.

例 2. 1-7 They thought that ...

6）学术论文在表达观点时往往是客观的，而不用主观的写法。

例 2. 1-8 This paper attempt to clarify ... 而不会写成：In this paper, I will attempt to ...

7）学术论文为了准确表达作者的观点，会用一些词汇准确表达论点的可靠程度，而不是简单地使用完全肯定或否定的句子。

例 2. 1-9 The possible reason may be that the suspensions have higher viscosity than that of pure water, especially at high particle volume fractions.

8）使用专业词汇。如例 2. 1-9 中 absorption application 指吸收式制冷的应用，thermal conductivity 指热导率，这些都是典型的空调制冷及能源领域的基础词汇，应掌握并应用在论文写作中，使论文严谨、规范。

例 2. 1-10 The objectives of this study are to evaluate the stability and to measure the thermal conductivity of binary nanoemulsions for absorption application.

练习

1. 读下面的句子，请分辨哪个句子更适合用在学术论文写作中。

1）The aim of this study is to find the mechanism to induce Marangoni convection.

2） We'll repeat these experiments this year.

3） These experiments are repeated.

4） We collected all data with a data acquisition instrument.

5） All data is collected with a data acquisition instrument.

2. 请将下面的句子补充完整。

1） This paper attempts to ________________________________.

2） One possible reason may be ________________________________.

3） This paper attempts to clarify ________________________________.

2.2 英文科技论文的写作技巧

2.2.1 范畴和分类的表达

为了清晰地表达观点，在论文的写作中经常使用分类。就像在日常生活中，功能良好的现代衣柜通常会设计出多个分格的空间，可以将衬衫、裤子、袜子、领带等分别放置于不同的空间。而老式的衣柜或箱子通常只有一个储物空间，所有不同种类的衣物都堆放在一个空间。显然现代衣柜由于进行了分类，可使人一目了然，迅速找到需要寻找的衣物。而老式衣柜或箱子却需要翻箱倒柜才能找到要找的衣服。

写作时也是一样，可把相似的东西可归为一类，在每一大类中又可划分为许多小类。这样有助于清晰地表达作者的观点。

例 2. 2-1　Air-cooling evaporators can be classified into the following three types: bare tube, plate-surface and finned-tube. The focus of this paper is on the finned-tube evaporator.

分类的常用句型如下：

1） There are three types （classes, categories, sorts, varieties） of … resulting from …

2） The effects are （may be, can be） classified … on the basis of （depending upon） …

3） The effects of … maybe （can be） grouped into three main categories.

另一种重要的表示分类的方法为用图表来表示分类。这种方法更加简洁且一目了然，如图 2. 2-1 所示。

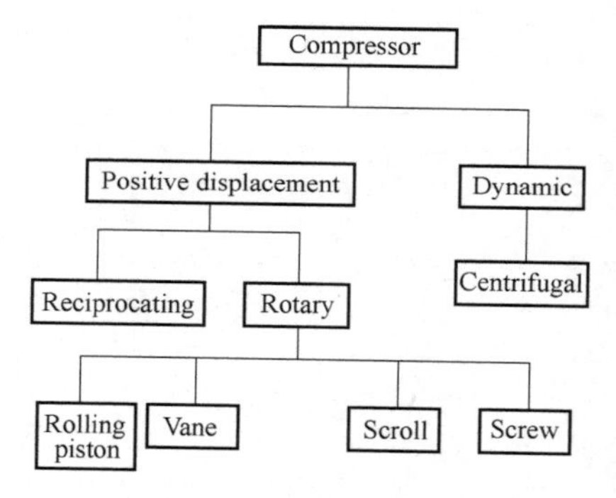

Fig. 2. 2-1　Classification of compressors

练习

1. 请用 2. 2. 1 节中给出的常用分类句型写句子，每个句型至少写出一个完整的句子。

2. 写一段文字将 2.2.1 节中的图 2.2-1 压缩机的分类用句子表达出来，并比较图表和文字表达的优缺点。

2.2.2 比较和对比的表达

对于两个以上的事物，为了清晰地阐述作者的观点，比较和对比在科技论文中经常出现。比较通常是对事物间相似之处进行比较，而对比是针对事物间不同之处的对比。

例 2.2-2 Let's buy this instrument, it's *cheaper than* that one but has the same measurement capacity.

该例句中 cheaper 是形容词 cheap 的比较级。单音节形容词的比较级一般是在形容词后加 er。双音节以 y 结尾的形容词，比较级是将 y 变为 i 再加 er，如 heavy-heavier，easy-easier，early-earlier。

双音节或更多音节的形容词，比较级是在形容词前加 more。例如，serious-more serious，expensive-more expensive，often-more often，difficult-more difficult，reliable-more reliable。

另外以 ly 结尾的副词前面也用 more 表示"更加……"。例如，more slowly，more seriously，more quietly，more carefully。

也有少量形容词，有不规则的比较级形式。例如，good-better，bad-worse，far-further。

在形容词比较级前还可使用 much，a lot，far，a bit，a little，slightly 等词。

例如：

Let's go by bus. It's much cheaper.

I feel much better than yesterday.

This instrument is much more expensive.

The problem is much more serious than we thought at first.

the ... the ... 越……越……

The younger you are, the easier a foreign language is to learn.

The warmer the weather, the better the patients feel.

表示比较（comparison），还有下面的用法。

1）The new test section is designed to be similar to the previous design but with a new liquid distributor.

2）Similar to the Fe_2O_3 nanoparticles, the vapor absorption rate significantly increases with increasing the concentration of CNT.

表示对比（contrast），请看例 2.2-3、例 2.2-4。

例 2.2-3 Even though the results of many studies have shown that the thermal conductivity and convective heat transfer of nanofluids are enhanced, the experimental results are quite different from each other because the heat transfer characteristics of nanofluids strongly depend on the thermal properties of the base fluids/nanoparticles, concentration of the nanoparticles and size/shape of the nanoparticles, etc.

例 2.2-4 The surface tension of aqueous lithium bromide increases with concentration. In contrast, the surface tension of aqueous lithium bromide containing 500 ppm n-octanol decrease with concentration.

表示比较的常用句型有：

A is like B（A and B are similar, A is similar to B, A resembles B）with respect to …

Both A and B …

A is as … as B.

A is no more … than B.

A … the same as B.

表示对比的常用句型有：

A is unlike B.

A differs from B.

A and B differ.

A is different from B.

A contrast with B.

A …, whereas（but, while）B …

B … more than A.

A is not as … as B.

表示比较的常用副词有：similarly, correspondingly, likewise。

表示对比的常用副词和短语有：on the other hand, in contrast, conversely。

使用比较和对比后，通常紧接着会给出评价和建议。

例 2.2-5 The absorption refrigeration systems have a lower COP and are more expensive than the vapor compression systems. Therefore, optimization from thermodynamic as well as economic points of view is needed.

练习

1. 完成句子，请用下列形容词的比较级形式填空。

reliable　serious　important　simple　good

1）My old car breaks down frequently recently. I need to buy a ____________ car.

2）The problem is ____________ than we thought at first.

3）The mechanism is ____________ than phenomena.

4）This design is ____________ than that one.

5）This is a ____________ suggestion to solve that problem.

2. 请用 much、far、a bit、slightly 等词填空。

1）Her illness was ____________ more serious than we thought.

2）It is ____________ more difficult to learn a foreign language outside the country where it is spoken.

3) You are driving too fast. Can you drive ______________ more slowly?

4) He has worked in England for many years, but his English is ______________ poorer than his brother who lives in Beijing.

3. 请用下列形容词的比较级填空。

expensive good little

1) The ______________ the hotel, the ______________ the service.

2) When you are on a business trip, the ______________ luggage you have the better.

4. 请用 in contrast, similarly, correspondingly, conversely 填空。

1) This device is expensive to buy. ______________, it is cheap to operate.

2) This device is expensive to buy. ______________, it is also expensive to operate.

5. 请写一段话，先使用对比，然后给出评价。

2.2.3 使用定义

虽然阅读学术论文的读者通常与作者为同一领域，对于专业领域内的术语，大家都熟悉，并不需要使用定义。但对于一些新出现的术语，或者只在小范围被少量专业人士了解的词汇，需要在论文中清晰地做出解释和说明，这时可以使用定义。

定义的常用表达方式见例 2.2-6~例 2.2-8。

例 2.2-6 Oscillating heat pipe is an active cooling device in which it converts intensive heat from a high-power generating device into kinetic energy of fluids to support oscillating motion.

例 2.2-7 Fluids with nanoparticles suspended in them are called nanofluids, a term proposed by Choi in 1995 at the Argonne National Laboratory, USA.

例 2.2-8 Wenzel roughness factor is defined as the ratio of the actual area of the solid surface to its projected area.

练习

1. 请为 air conditioner 写定义。

2. 请为 refrigerator 写定义。

3. 请为 water chiller 写定义。

4. 请为 fan coil 写定义。

2.2.4 怎样写概括

学术论文每个段落的起始，经常用一句话概括本段的主要内容。后续的内容从概括展开，可以用例子、分类、定义等支持段首的观点。做概括时，如果不是百分百肯定，不要使用绝对肯定的语气，而要根据肯定的程度，恰当地用词汇表达出确定程度，这也是学术论文严谨性的一个方面。

表示百分百肯定或否定的常用词有：all，every，each，no，none，not any；always，never；definite，certain，undoubted，clear；definitely，undoubtedly，certainly，clearly；will，will not，is，are，must，have to。

表示较大程度的肯定的常用词汇有：a majority of，many，much；usually，normally，generally，as a rule，on the whole，often，frequently；probable，likely；presumably，probably；should，would，ought to。

下面的词汇肯定程度次之：some，several，a number of；sometimes，occasionally；uncertain；possible，possibly，perhaps，maybe；can，cannot，could，could not，may，may not；might，might not。

下面的词汇肯定程度更低：a minority，a few，a little，few，little；rarely，seldom，hardly ever，scarcely ever；unlikely，improbable。

有一些动词也可以表示出某种程度的不确定性，弱化语气，例如：to seem，to appear，to believe，to assume，to suggest，to speculate，to estimate，to tend，to think，to argue，to indicate，to project，to forecast。

有些词可以加强语气，例如：complete，very，full，thorough，total，extreme，absolute，definite，great，deep，strong，high；completely，fully，thoroughly，totally，entirely，absolutely，definitely，greatly，deeply，strongly，highly。

例 2.2-9 It is possible to reduce this error by adjusting the superheat.

例 2.2-10　In the following, the operating conditions and the main results gathered during the experiments are highlighted and deeply analyzed.

例 2.2-11　This conclusion seems highly questionable because...

例 2.2-12　This concept seems hardly applicable as an efficient industrial drying technology.

练习

1. 请用 seem, appear, believe, assume, suggest, speculate, estimate, tend, think, argue, indicate, project, forecast 改写下列句子以降低肯定程度。

1) Scientists say that the population of China will decline by 10 millions each year in the next 30 years.

2) When a region is involved in a war, the inhabitants will migrate.

2. 请使用 completely, fully, thoroughly, totally, absolutely, definitely, deeply, strongly, highly 或 complete, full, thorough, total, extreme, absolute, definite, great, deep, strong, high 等词汇改写下列句子以增强肯定程度。

1) Scientists agree about the mechanism he proposed.

2) There is a disagreement between the two sections of the research community.

3) The statement issued by the research team is doubtful.

3. 请改写下面的句子以降低肯定程度。

The experimental results show the heat and mass transfer are enhanced obviously. The reason of enhancement is caused completely by adding additives in solution.

2.2.5　如何描述过程

在科技论文的写作中往往需要描述实验过程，这些过程往往是一系列事件顺序发生

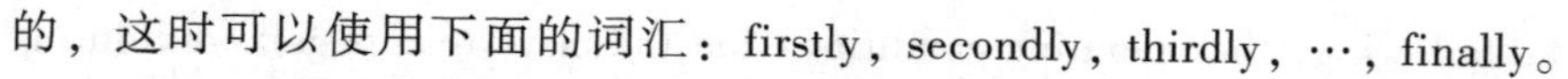

的，这时可以使用下面的词汇：firstly，secondly，thirdly，…，finally。

或者用其他表示时间的词，如：then，at that time，next year，at the beginning of，five years ago，in（year）；until，by，before，up to that time，in the weeks leading up to，prior to；in the meantime，at the very moment，during，simultaneously；subsequently，after，afterwards，then，next，presently，after a while，later，eventually，finally，at last 等

练习

1. 请将下面一段文字翻译为英语。

先用电子天平配制溶液样本，然后取 12 克溶液注入表面皿，将表面皿放入吸收器内的吊篮内，盖上吸收器盖，拧紧吸收器盖上的螺栓，起动真空泵，对整个实验系统抽真空，待真空表显示-1atm，关闭真空泵，关闭吸收器和蒸气发生器的气体进出口的阀门。起动恒温水浴，设定恒温水浴温度为理想值，30 分钟后，打开所有阀门，起动数据采集系统，蒸气从蒸气发生器进入吸收器，吸收实验开始，5min 后，关闭所有阀门及数据采集系统，导出数据采集系统记录的数据，吸收实验结束。

2.2.6 动词的名词化

在英文科技论文的写作中，经常使用动词的名词化形式，而不是直接使用动词。

例如：application 在论文中出现的概率会大于 apply。

常见的动词的名词化形式有：enhancement，requirement，introduction，arrangement，establishment，installation，ventilation，simulation，consideration，absorption，operation，acquisition，extraction，information，relation，duration，consumption，distribution，reduction，evaluation，indication，investigation，assessment，agreement，measurement，combination，comparison 等。

例 2.2-13 No direct calculations of the thermal coefficient of performance were possible due to these problems.

例 2.2-14 Viscosity measurements were performed using a viscometer.

练习

1. 请分别用 enhancement、consumption、measurement、investigation 造句。

2.2.7　观点的表达

在科技论文中，作者需要针对某一问题提出自己的观点，并在文中提供证据支持自己的观点。表达自己的观点时常用下面的词汇、短语和句子：suggest，present，propose，point out。

例 2.2-15　Binary nanoemulsions, oil-droplet suspensions in binary solution (H_2O/LiBr), were proposed to enhance the heat and mass transfer performance of absorption refrigeration systems.

例 2.2-16　Many mechanisms such as the ballistic nature of heat transport in nanoparticles, the effects of nanoparticle clustering, the Brownian motion of the particles, and the layering of liquid molecules at the particle-liquid interface have been suggested to be responsible for such anomalous enhancement.

练习

1. 请针对专业领域内的某一问题，写一段话表达自己的观点。

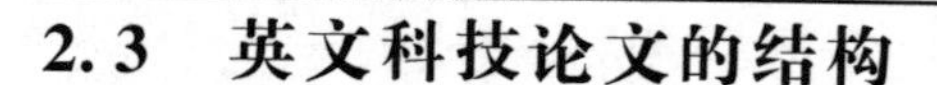

2.3 英文科技论文的结构

一般专业期刊上的原创论文，包含论文标题、作者及工作单位、关键词、摘要、介绍、术语表、研究方法、结果和讨论、结论、致谢和参考文献等部分。

论文的标题要能够尽量多地反映论文的核心内容并尽量使题目不要过长，题目中一般不包含冠词。提供的关键词一般需五个左右，是为满足文献检索的需要，为了让读者能方便地通过输入关键词找到你的论文。因此，关键词必须反映论文的核心内容。

摘要一般不超过几百字（words）。摘要需要简洁，简要介绍研究目的、研究方法、研究结果和主要结论。摘要是独立于论文主体的，因此要避免在摘要中引用参考文献，也要避免在摘要中使用非标准或不常用的缩写。用美国工程索引 EI（the engineering index）等文献数据库搜索时，论文的题目和摘要都可直接搜索到，而论文全文只有在读者所在的单位购买了期刊或论文集的论文全文资源时，全文才能看到。

论文的主体部分为介绍、研究方法、结果和讨论，即 IMRD（introduction, method, results and discussion）。论文主体中最重要的部分是结果和讨论，是科技论文最核心的部分，应下大力气写到最好。结论也是论文中必须写的部分，作者将研究得到的结论在这里再做一次总结，方便读者阅读。论文中提到的他人研究、观点需要在文中用带方括号的数字做索引，索引序号按在文中出现的顺序编号，对应地在参考文献中需要注出文献出处，也可按作者姓氏的首字母排列顺序做索引。受某基金资助或得到某人的帮助需要在致谢部分表示感谢。

2.3.1 怎样写介绍部分

在介绍部分需要划定作者研究的范围，回顾国内外此研究范围的研究状况、研究热点及指出研究空白在哪里，本研究针对哪个研究热点或空白展开。介绍部分不要简单地罗列很多文献的研究内容，要选择与本论文研究内容密切相关的文献，按照一定的逻辑顺序，概括总结其他同行都研究了哪些内容，得出了哪些结论，哪些方面还需要深入研究，随后引入本论文的研究内容及研究方法。

2.3.2 怎样写研究方法部分

实验研究论文通常在研究方法部分介绍实验系统、实验步骤、数据处理方法、测量技术、不确定度分析等。数值模拟论文通常要在这一部分介绍控制体的选择、守恒方程、离散化方法、初始条件及边界条件等。为了更清晰地说明实验系统、测量技术及控制体的质量及能量流入流出情况，通常需要画出实验系统、测量系统原理图及控制体的简图。论文中出现的图表，要按在文中出现的顺序编号，表要有表名，图要有图名。

2.3.3　怎样写结果和讨论部分

实验研究或数值研究的结果要用图线在结果和讨论部分列出。从实验或数值计算的结果可以发现什么，得到什么结论，分析现象产生的可能原因是什么，作者的观点是什么，与他人的研究结果对比有哪些异同，为什么会出现差异等都要在结果和讨论部分写出，这个部分是论文中最重要的部分。

2.4 科技论文的写作原则

科技论文的写作需要遵循清晰性（clarity）、诚实性（honesty）、现实性（reality）、相关性（relevance）的原则。

清晰性原则是指作者需要准确、清楚地阐明论文中提到的一切，使读者读论文时不会产生模糊或不清晰的印象。前面讲过的定义、比较、对比、分类、专业词汇的应用等都有助于提高论文写作的清晰性。

诚实性是指论文作者讲的话必须是有证据支持的，因此论文中表达的观点要实事求是，不可以夸大。作者提出的观点如果不是百分之百肯定或否定，就不能使用 absolutely，definitely，all，each，never 等词。使用部分肯定的词汇时，也要分清常用词汇表示肯定程度的区别，恰当地使用，例如使用 usually 就比 sometimes 表示事物出现的频率高。如果作者提出了一个观点来解释一个现象，如果不是百分之百肯定就是这个原因引起的，就必须使用 possible 这个词以准确地表达肯定程度。

现实性是指论文的作者需要清楚论文的读者对于专业领域的哪些内容是熟知的，哪些是不了解的。这样写论文时有些内容就不需要解释和说明，而有些内容则需要使用定义等做清晰的说明。

相关性是指作者要时刻提醒自己，写的内容不要偏离主题，所有的内容是相关的，都是为表达主题服务的，不要写与主题无关的内容。

2.5　出版道德

在同行评议期刊上发表论文是构成令人尊敬的知识网络的基石，出版行为的所有参与者包括作者、期刊编辑、同行评议人、出版机构、主办期刊的机构都需遵循出版的道德规范。了解关于研究和出版的道德规范可以使年轻的科研人员避免不当行为。了解科学研究和出版的道德边界可以使年轻人从一个最好的起点出发，未来有很好的发展。科学研究中的不当行为和出版中违背道德规范可能是故意的也可能不是故意的，要尽量避免。

常见的科学研究中的不当行为包括伪造实验数据，篡改数据、图像、流程等。常见的违反出版道德的行为包括同时向两家或两家以上的期刊投递同一篇论文，向另一家期刊投递以前发表过的论文或部分内容并且没有注明已发表过的事实，从不同角度对同一研究进行阐述未引用原始论文，向期刊投递其他语种已发表的论文且未提供原论文的信息等。

2.6 如何避免科学研究和出版中的抄袭行为

抄袭或剽窃（plagiarism）是学术上的不诚实或偷窃行为。当一个人抄袭了其他人论文中的句子、段落，或将其他人的观点想法作为自己提出的想法发表就是一种抄袭行为。

下面列出几种最常见的学术抄袭行为。

1）直接大段抄袭已经出版的书、期刊、学位论文等的内容写在自己的论文中，不使用引用，不给出出处，让人觉得这些内容是你自己写的。

2）直接抄袭已经出版的书、期刊、学位论文等的句子，不使用引用，不给出出处，让人觉得这些内容是你自己写的。

3）抄袭别人的数据和图表及研究结果，当作自己的研究成果。

4）将集体的研究成果当成自己个人的研究成果。

5）用自己的话重述别人的观点，没有给出引用和出处。

怎样在撰写学术论文时避免抄袭呢?

1）直接引用别人论文中的句子时，要使用引号。

例 2.6-1 Chan Woo Park reported, "For the bare tube, absorption rate with heat transfer additive is enhanced as high as 3.76 times of that without the heat transfer additive."[1]

2）用自己的语言重述或总结作者的观点并注明出处时，不使用引号。

例 2.6-2 Hozawa et al. reported that the presence of an island of surfactant is not a necessary condition to initiate Marangoni convection, but it can provide more violent convection for a longer time by acting as a reservoir of the surfactant on the surface [6].

相应地在参考文献部分要按一定的著录规则给出文献出处。

[6] HOZAWA M, INOUE M, SATO J, et al. Marangoni convection during steam absorption into aqueous LiBr solution with surfactant. J Chemical Engineering of Japan, 1991, 24(2): 209-14.

练习

1. 请将下面一段直接引语用自己的语言重述，总结作者的观点以便用在自己的论文中，并给出出处。

Chan Woo Park reported, "For the bare tube, absorption rate with heat transfer additive is enhanced as high as 3.76 times of that without the heat transfer additive."[1]

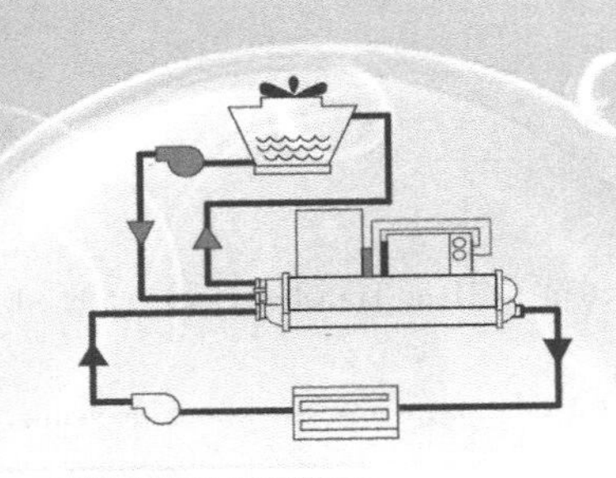

附录

制冷空调领域主要的国际期刊介绍

1. International Journal of Heat and Mass Transfer

《国际传热传质》期刊是传热传质学领域著名的国际期刊，办刊目的是为传热、传质学领域的研究人员及工程师提供交流的平台。该刊侧重分析研究、实验研究，并强调论文需有助于对传递过程的基本理解和在应用中有助于工程问题的解决，主要面向学术机构和工业领域的研究人员、传热和热力系统的工程师、化工及相关领域的过程工程师。论文类型包括原创研究论文、研究综述、短的通讯稿、书信及给编辑的信等。

刊发的论文主要包括以下三方面内容：

1）New methods of measuring and/or correlating transport-property data（测量的新方法和热质传递特性数据关联的新方法）。

2）Energy engineering（能源工程）。

3）Environmental applications of heat and/or mass transfer（传热传质的环境应用）。

网址为：

http://www.journals.elsevier.com/international-journal-of-heat-and-mass-transfer。

2. International Journal of Refrigeration

《国际制冷》期刊是制冷领域著名的国际期刊。由国际制冷学会（IIR）主办，Elsevier出版。期刊覆盖制冷领域的理论和实践，包括热泵、空气调节和食品的储藏和运输，面向希望了解制冷、空调及相关领域的最新研究进展及工业技术发展的读者。期刊已出版多期关于制冷剂替代、沸腾、冷凝、热泵、食品制冷、二氧化碳制冷、氨制冷、烃制冷、磁制冷、吸收式制冷、相变材料、冰浆等方面的专刊。除了发表原创研究论文外，期刊还发表研究综述论文、IIR主办的国际制冷大会上提交的论文、短的研究报告、描述初步研究结果和实验细节的书信、给编辑写的关于近期研究的讨论和争议的书信，还包括近期的活动安排、会议报告和书评等方面内容。

网址为：

http://www.journals.elsevier.com/international-journal-of-refrigeration

参 考 文 献

[1] INCROPERA F P, DEWITT D P, BERGMAN T L, et al. Introduction to Heat Transfer [M]. 5th ed. Hoboken, NJ: JOHN WILEY & SONS, 2007.

[2] Behler-Young Company. Accumulator or Receiver [EB/OL]. [2019-1-6]. http://www.behler-young.com/tech-tips/refrigeration-tips/accumulator-or-receiver.

[3] STOECKER W F, Jones J W. Refrigeration and air conditioning [M]. 2nd ed. New York: McGraw-Hill Book Company, 1982.

[4] HEROLD K E, RADERMACHER R, KLEIN S A. Absorption Chillers and Heat Pumps [M]. Boca Raton, FL: CRC PRESS, 1996.

[5] DOSSAT R J, HORAN T J. Principles of Refrigeration [M]. 5th ed. Upper Saddle River, New Jersey: Prentice Hall, 2002.

[6] Carrier Corporation Technical Training. Distribution Systems-Water Piping and Pumps [Z]. Syracuse: Carrier Corporation, 2005.

[7] AMEEN A. Refrigeration and Air Conditioning [M]. Delhi: Prentice Hall of India, 2006.

[8] BAILEY S. Academic Writing-A handbook for International Students [M]. 2nd ed. Abingdon, Oxon: Routledge, 2006.

[9] CHADDERTON D V. Air Conditioning: A Practical Introduction [M]. London: E & FN Spon, 1997.

[10] CARTER R, MCCARTHY M. Cambridge Grammar of English [M]. Cambridge: CAMBRIDGE UNIVERSITY PRESS, 2006.

[11] HAMP-LYONS L, HEASLEY B. Study Writing: A Course in Writing Skills for Academic Purposes [M]. 2nd ed. Cambridge: CAMBRIDGE UNIVERSITY PRESS, 2006.

[12] HEWINGS M. Advanced Grammar in Use [M]. 2nd ed. Cambridge: CAMBRIDGE UNIVERSITY PRESS, 2005.

[13] HUNDY G F, TROTT A R, WELCH T C. Refrigeration and Air-Conditioning [M]. 4th ed. Amsterdam: Butterworth-Heinemann/Elsevier, 2008.

[14] LANGLEY B C. Heat Pump Technology [M]. 3rd ed. New Jersey: Prentice Hall, 2002.

[15] MUNSON B R, YOUNG D F, OKIISHI, T H. Fundamentals of Fluid Mechanics [M]. 5th ed. Hoboken, NJ: JOHN WILEY & SONS, 2006.

[16] MONTGOMERY R, MCDOWALL R. Fundamental of HVAC Control Systems [M]. Atlanta, GA: ELSVIER, 2008.

[17] MORAN M J, SHAPIRO H N. Fundamentals of Engineering Thermodynamics [M]. 5th ed. Hoboken, NJ: JOHN WILEY & SONS, 2004.

[18] MURPHY, R. English Grammar in Use [M]. 3rd ed. Cambridge: CAMBRIDGE UNIVERSITY PRESS, 2004.

[19] WELTY J R, WICKS C E, WILSON R E, et al. Fundamentals of Momentum, Heat and Mass Transfer [M]. 4th ed. Singapore: JOHN WILEY & SONS (Asia), 2005.

[20] 朱明善，刘颖，林兆庄，等. 工程热力学 [M]. 北京：清华大学出版社. 1995.

[21] 廖道平，吴业正. 制冷压缩机 [M]. 北京：机械工业出版社，2001.

[22] 吴业正，朱瑞琪，曹小林，等. 制冷原理及设备 [M]. 3 版. 西安：西安交通大学出版社，2010.

[23] 赵荣义，范存养，薛殿华. 空气调节 [M]. 4 版. 北京：中国建筑工业出版社，2008.

[24] 陈芝九，吴静怡. 制冷装置自动化 [M]. 2 版. 北京：机械工业出版社，2010.

[25] 张道真. 实用英语语法 [M]. 北京：外语教学与研究出版社，1995.